PAH
in
Work Atmospheres:
Occurrence
and
Determination

Editors

Alf Bjørseth, Dr. phil.
Director
Petroleum Research Center
Norsk Hydro
Bergen, Norway

Georg Becher, Dr.rer.nat.
Head, Chemical Laboratory of
Toxicology
National Institute of Public Health
Oslo, Norway

CRC Press
Taylor & Francis Group
Boca Raton London New York

CRC Press is an imprint of the
Taylor & Francis Group, an **informa** business

First published 1986 by CRC Press
Taylor & Francis Group
6000 Broken Sound Parkway NW, Suite 300
Boca Raton, FL 33487-2742

Reissued 2018 by CRC Press

© 1986 by CRC Press, Inc.
CRC Press is an imprint of Taylor & Francis Group, an Informa business

No claim to original U.S. Government works

Library of Congress Cataloging-in-Publication Data

Bjørseth, Alf.
    PAH in work atmospheres

    Bibliography: p.
    Includes index.
    1. Polycyclic aromatic hydrocarbons--Toxicology.
2. Polycyclic aromatic hydrocarbons--Environmental
aspects. 3. Polycyclic aromatic hydrocarbons--Metabolism.
4. Polycyclic aromatic hydrocarbons--Measurement.
5. Industrial toxicology. I. Becher, Georg. II. Title.
RA1242.P73B56   1986   616.99'4'071   85-30731
ISBN 0-8493-6064-1

A Library of Congress record exists under LC control number: 85030731

Publisher's Note
The publisher has gone to great lengths to ensure the quality of this reprint but points out that some imperfections in the original copies may be apparent.

Disclaimer
The publisher has made every effort to trace copyright holders and welcomes correspondence from those they have been unable to contact.

ISBN 13: 978-1-315-89621-2 (hbk)
ISBN 13: 978-1-351-07531-2 (ebk)

Visit the Taylor & Francis Web site at http://www.taylorandfrancis.com and the
CRC Press Web site at http://www.crcpress.com

PREFACE

Polycyclic aromatic hydrocarbons (PAH) are a group of chemical compounds that are strongly suspected of exhibiting carcinogenic effects in humans. This suspicion is based on long-term observation, i.e., in animal bioassays, occupational medicine, and epidemiology. Such observations have been made for a long time. The earliest association of a human cancer with exposure to an environmental hazard was made more than two centuries ago by Sir Percivall Pott who reported an increased incidence of scrotal cancer among chimney sweeps. The coal soot, suspected to be responsible for this scrotal cancer, was not known until Kennaway and his colleagues reported their work in the 1930s. Since that time, ample evidence has been generated showing that increased industrialization and man's use of fossil fuels result in a wide distribution of PAH in the environment. Particularly high levels of PAH are to be found in occupational environments.

This book deals with the sources, distribution, analytical methods, and monitoring of PAH in the occupational environment. It is our hope that the publication of this book will make a contribution to understanding the formation and determination of PAH in work atmosphere and that it will make a particular contribution to occupational health projects.

Much of the information given in this book has been generated in studies carried out in cooperation with the Norwegian aluminum industry. The financial contribution to this work from the Nordic Aluminum Industry's Secretariat for Health, Environment, and Safety is gratefully acknowledged.

The authors would also like to acknowledge the inspiring working conditions provided by the colleagues at the Center for Industrial Research in Oslo.

THE EDITORS

Alf Bjørseth, Dr. phil., graduated from the University of Oslo in 1969 with a background in mathematics, physics, and chemistry with particular emphasis on physical chemistry. He finished his doctorate in 1979 in Environmental Chemistry.

Dr. Bjørseth is currently Director of Norsk Hydro's Petroleum Research Center in Bergen, Norway. His previous engagements include Research Manager in the Chemistry Division of the Central Institute for Industrial Research in Oslo, Norway and Associate Section Manager at Battelle Columbus Laboratories in Columbus, Ohio. He is also co-founder of the Scandinavian Center for Advanced Technology (SCATEC). Dr. Bjørseth's research interests include studies of polycyclic aromatic hydrocarbons and their transformation products — with emphasis on the application of high resolution gas and liquid chromatographic techniques. Dr. Bjørseth has published over 60 papers in these areas, authored and edited several books, given a number of lectures in the U.S. and Europe, and chaired international symposia devoted to polycyclic aromatic hydrocarbons.

He is a member of the Norwegian Chemical Society and Norwegian Petroleum Society where he is currently serving as President of the Bergen Division.

Georg Becher, Dr.rer.nat., is Head of the Chemical Laboratory in the Department of Toxicology at the National Institute of Public Health in Oslo, Norway. Dr. Becher obtained his M.S. and Ph.D. degrees in 1970 and 1974, respectively, from the Department of Organic Chemistry, University of Gottingen, West Germany. He did his postdoctoral work at the Department of Chemistry, University of Oslo, Norway and served as an Assistant Professor at the Department of Chemistry, University of Regensburg, West Germany. In 1979, he joined the Environmental Chemistry Department at the Central Institute for Industrial Research in Oslo as a research scientist. Since 1983, Dr. Becher is in his present assignment at the National Institute of Public Health in Oslo.

Dr. Becher has been a recipient of the postgraduate fellowship of the Funds of Chemical Industry, West Germany and is a member of the German and the Norwegian Chemical Society. Dr. Becher is currently engaged in the analyses of trace quantities of potentially hazardous compounds in environmental samples. Among others, he is interested in the development of new methods for determination of human exposure to PAH.

TABLE OF CONTENTS

Chapter 1

PAH AS OCCUPATIONAL CARCINOGENS

I. HISTORICAL REVIEW

The history of the detection, synthesis, and biological study of polycyclic aromatic hydrocarbons (PAH) is closely related to the establishment of the carcinogenicity of coal tar. Already in 1775, the British surgeon Sir Percivall Pott published studies which indicated increased incidences of scrotal cancer among chimney sweeps in England, resulting from prolonged contact of the skin with coal soot.[1] These chimney sweeps were, as children, forced to climb the narrow English chimneys and remove soot from the inside of the chimney flue. Dr. Pott assumed that the cause of the cancer development was soot and tar. It is, in retrospect, interesting to note that Dr. Pott did not regard scrotal cancer as a serious problem, since it could be removed without the least inconvenience. This part of Dr. Pott's observations is probably not valid today.

About 100 years later, high incidences of skin cancer were reported among workers in the paraffin refining,[2] shale oil,[3] and coal tar industries.[4] However, early attempts to produce cancer in experimental animals with the raw materials of these industries were unsuccessful. It was not until 1915 that the Japanese pathologist Yamagiwa and his colleagues at the Imperial University of Tokyo[5] succeeded in inducing tumors with coal tar. Professor Yamagiwa and his assistant Ichikawa showed[5] that if they repeatedly painted the inside of the ears of rabbits (two to three times a week), malignant tumors developed at the site of the skin painting after 3 months, and these later proved to be cancer tumors which grew over the whole ear. This was the first time that cancer tumors were produced experimentally by means of chemicals. It is interesting to note that Yamagiwa and Ichikawa made a very good choice by painting the ears of rabbits, since these show very high sensitivity toward experimental carcinogens. If they had painted the skin of dogs or guinea pigs instead of rabbits, they probably would not have found any tumor development. Soon thereafter, Tsutsui[6] obtained malignant skin tumors in mice also by painting them with coal tar. This was the first experiment with a design that later should prove to be one of the most-used animal models for studies of carcinogenic compounds: skin painting on mice. Following Tsutsui's observation, other scientists showed that soot could be extracted and that these extracts also produced skin cancer on mice. Therefore, by 1925 it was well established that coal, tar, and soot contained material that was carcinogenic both to animals and to man.

In the 19th century, a discovery was also made which was of basic importance to the chemical understanding of PAH. The recognition in 1865 that carbon atoms could be linked in closed rings was one of the great conceptual advances of chemistry. The discovery was made by August Kekulé.[7] It is said that he solved the riddle of the structure of the benzene molecule after having a dream in which dancing snakes bit their own tails. From the understanding of the structure of the benzene molecule it was only a short step to the recognition that the naphthalene molecule has two fused benzene rings. The structure of the multiple-ring, or polycyclic compounds will be discussed below.

Soon after Kekulé's discovery, the classical paper of Berthelot was published[8] in which he showed that the heating of acetylene formed a tar which contained benzene and other aromatic hydrocarbons. A likely source of the PAH was therefore recognized.

After the animal studies had demonstrated the presence of carcinogenic compounds in coal tar, studies were undertaken to attempt to characterize the carcinogens in this material. All that was known then was that the carcinogenic compounds were concentrated in the high-boiling fractions of the tar. Kennaway and his group at the (then) Research Institute of the Cancer Hospital in London (subsequently the Chester Beatty Research Institute and now the Institute of Cancer Research) started in 1922 a series of chemical and physical studies of coal tar. After years of studies of various synthetic tars and synthesized PAH, they found that the fluorescence spectrum of benz(a)anthracene was very similar to those of the carcinogenic tars, although shifted to longer wavelengths.[9,10] This suggested that the carcinogens in coal tar contained a benz(a)anthracene nucleus with some additional substituents. They then synthesized dibenz(a,h)anthracene and its 3-methyl-derivative. The fluorescence spectra of the two hydrocarbons showed the same characteristic bands as benz(a)anthracene and the tars, now at intermediate wavelengths. When these compounds were tested by skin painting of mice, they were found to produce tumors.[10] This was the first recognition of the carcinogenic activity of a chemical of defined structure.

Shortly thereafter, Hieger and co-workers[11] initiated a large-scale isolation of the carcinogen(s) in coal tar. Starting with two tons of pitch, they performed a series of fractional distillations, differential extractions, and crystallizations. The various fractions were characterized by their fluorescence spectra and their carcinogenic activity by skin painting on mice. In the course of about 2 years they had isolated about 7 g of a yellow crystalline material which had the correct fluorescence spectrum and exhibited high carcinogenic activity. This material was found to consist of two isomeric compounds, benzo(a)pyrene (BaP) and benzo(e)pyrene (BeP). Both the synthetic and the isolated samples of BaP were highly carcinogenic.[11]

By 1930, the English research group had managed to isolate dibenz(a,h)anthracene and BaP, both of which proved to be strongly carcinogenic in skin-painting experiments, and BeP which did not show carcinogenic properties. In the years to follow, the English researchers isolated 60 new compounds and performed in one period 146 different skin-painting experiments simultaneously. During these years, a large number of PAH compounds were isolated and synthesized, and were tested for carcinogenic effects by skin painting.

These studies were the first of a large number of studies of the carcinogenic properties of PAH. By 1976, more than 30 parent PAH compounds and several hundred alkyl derivates of PAH were reported to have some carcinogenic effects.[12,13] This makes PAH and their derivatives the largest single class of chemical carcinogens known today. Recently, the International Agency for Research on Cancer (IARC) made a critical evaluation of all available data on the carcinogenic effect of 32 PAH.[14] It was concluded that 11 PAH were found to be proven carcinogens in experimental animals, 6 were found likely to be carcinogens, 12 had inconclusive evidence, and 3 were proven noncarcinogens. The results from the IARC evaluation are summarized in Table 1.

II. EPIDEMIOLOGICAL EVIDENCES FOR CANCER RISK FROM PAH

To the best of our knowledge, there are no reports on epidemiological studies where occupational exposure was limited exclusively to PAH. However, there are a large number of epidemiological studies of occupational exposure in industries where the work atmosphere contained PAH in addition to other pollutants. The results from many of these studies indicate that PAH is an occupational hazard and, combined with the fact that many PAH are shown to be animal carcinogens, form the basis for the interest in characterizing PAH in the work environment. In the following, we will briefly summarize the results of some epidemiological studies from occupations and industries known to have high PAH exposure.

A. Chimney Sweeps

As already discussed, Sir Percival Pott published a classical report in 1775 on an increased frequency of scrotal cancer among chimney sweeps.[1] Other studies related to the same occupation are rare. In 1937, Henry published a study on the relation of cutaneous cancer

3

Table 1

EVALUATION OF THE CARCINOGENIC ACTIVITY OF SELECTED PAH

Compound	Evidence of carcinogenicity in experimental animals[a]	Evidence of activity in short-term tests[b]	Mutagenicity to *Salmonella typhimurium* (Ames test)[c]
Fluorene	I	I	Neg
Phenanthrene	I	L	Pos
Anthracene	No	No	Neg
Fluoranthene	No	L	Pos
Pyrene	No	L	Pos
Benzo(a)fluorene	I	I	Inadequate
Benzo(b)fluorene	I	I	Conflicting
Benzo(c)fluorene	I	I	Inconclusive
Benzo(ghi)fluoranthene	I	I	Pos
Benzo(c)phenanthrene	I	I	Pos
Cyclopenta(cd)pyrene	L	S	Pos
Benz(a)anthracene	S	S	Pos
Chrysene	L	L	Pos
Triphenylene	I	I	Pos
Benzo(b)fluoranthene	S	I	Pos
Benzo(j)fluoranthene	S	I	Pos
Benzo(k)fluoranthene	S	I	Pos
Benzo(e)pyrene	I	L	Pos
Benzo(a)pyrene	S	S	Pos
Perylene	I	I	Pos
Indeno(1,2,3-cd)pyrene	S	I	Pos
Dibenz(a,c)anthracene	L	S	Pos
Dibenz(a,h)anthracene	S	S	Pos
Dibenz(a,j)anthracene	L	I	Pos
Benzo(ghi)perylene	I	I	Pos
Anthanthrene	L	I	Pos
Coronene	I	I	Pos
Dibenzo(a,e)fluoranthene	L	No	No
Dibenzo(a,e)pyrene	S	I	Pos
Dibenzo(a,h)pyrene	S	I	Pos
Dibenzo(a,i)pyrene	S	I	Pos
Dibenzo(a,l)pyrene	S	No	No

[a] Overall evidence of carcinogenicity in experimental animals. S = *Sufficient evidence:* there is an increased incidence of malignant tumors (a) in multiple species or strains; (b) in multiple experiments; or (c) to an unusual degree with regard to incidence, site, or type of tumor, or age at onset. Additional evidence may be provided by data on dose-response effects as well as information from short-term tests, or on chemical structure. L = *Limited evidence:* the data suggest a carcinogenic effect but are limited because (a) the studies involve a single species, strain, or experiment; (b) the experiments are restricted by inadequate dosage levels, inadequate duration of exposure to the agent, inadequate period of follow-up, poor survival, too few animals, or inadequate reporting; or (c) the neoplasms produced often occur spontaneously and, in the past, have been difficult to classify as malignant by histological criteria alone. I = *Inadequate evidence:* the studies cannot be interpreted as showing either the presence or absence of a carcinogenic effect because of major qualitative or quantitative limitations. No = *No evidence:* this applies when several adequate studies show, within the limits of the tests used, that the chemical is not carcinogenic. The number of negative studies is small since, in general, studies that show no effect are less likely to be published than those suggesting carcinogenicity.

[b] Overall evidence of activity in short-term tests. S = *Sufficient evidence:* there were a total of at least three positive results in at least two of three test systems measuring DNA damage, mutagenicity, or chromosomal anomalies. When two of the positive results were for the same biological endpoint, they had to be derived from systems of different biological complexity. L = *Limited evidence:* there were at least two positive results, either for different endpoints or in systems representing two levels of biological complexity. I = *Inadequate evidence:* there were too few data for an adequate evaluation, or there were contradictory data. No = *No evidence:* there were many negative results from a variety of short-term tests with different endpoints, and at different levels of biological complexity. If certain biological endpoints are not adequately covered, this is indicated.

[c] Neg = No mutagenic activity was observed in the presence or absence of an exogenous metabolic system. Pos = The compound was mutagenic in the presence of an exogenous metabolic system. No = No data available.

to occupation in England and Wales for the period 1911 to 1935.[15] In this period, a total of 147 fatal cases of cutaneous cancer in chimney sweeps was recorded, of which 70% were cancers of the scrotum. The crude death rate for cancer of the scrotum for the number of sweeps at risk was estimated to be 620 per million. Henry noted that this was 98 times higher than that for the general population. Kupetz[16] reported a double risk for pulmonary carcinoma among chimney sweeps in Berlin, and Hansen[17] found an increased risk of cancer among Danish chimney sweeps. Recently, Hogstedt et al.[18] published a cohort study on mortality among all members (2971) of the Swedish Chimney Sweeps Union who were active in 1950 or later and have been members for at least 10 years. The observed numbers of deaths before the age of 80 were compared with sex-, calendar year-, and age class-specific expectancy values from the national statistics of 1951 to 1979. Observed deaths were 230 vs. 197.6 expected. This result was due to a significant excess of deaths from tumors, particularly lung and esophageal cancer, and from nonmalignant chronic respiratory diseases. The multifold increased risk from these diseases could hardly be explained by extreme smoking or alcohol habits, but rather by exposure to PAH, nitrogen compounds, arsenic, and asbestos in combination with exposure to sulfur dioxide.

B. Aluminum Workers

That there might be an occupational hazard of cancer in the aluminum industry was realized by Kreyberg[19] in 1959, when he drew attention to the presence of BaP and other PAH in the air of potrooms. BaP, he believed, was likely to have been a cause of excess mortality from lung cancer in makers of coal gas, and also, at least in part, might have caused the excess mortality from lung cancer that was commonly found in urban dwellers compared to rural. It seemed logical, therefore, that it should imply a specific hazard in the primary aluminum industry.

However, it was not until 1971 that Konstantinov and Kuz'minykh published the first report suggesting that aluminum production workers did in fact suffer from an excess of lung cancer and probably skin cancer.[20] Since then, many other papers have been published which have variously suggested that aluminum workers might also suffer from a wide range of other cancers or, conversely, that they might suffer no unusual hazard of cancer at all.[21]

Three cohort studies comprise the great majority of the epidemiological material that is of substantial value in determining the reality and extent of any specific cancer hazard in the primary aluminum industry. These studies were performed by Gibbs and Horowitz[22] for Quebec, Canada, by Andersen et al.[23] for Norway, and by Rockette and Arena[24] for the U.S. The Canadian study[22] reported on lung cancer mortality of 5406 men (cohort 1) employed at an aluminum smelter on January 1, 1950, and of 485 men employed at a second plant (cohort 2) on January 1, 1951. For each man, the total number of years of exposure to tars, number of years since first exposure to tars, and an index of the degree of exposure expressed in tar-years were calculated. More than 99% of the men in the first cohort and just less than 99% of the men in the second cohort were traced.

The results showed that there was a definite dose-response relationship between lung cancer mortality and tar-years and years of exposure. The standardized mortality ratio (SMR) for persons exposed for more than 21 years to the higher levels of tars was 2.3 times that of persons not exposed to tars. Although smoking may still be a factor, the evidence suggests that the increased risk of lung cancer is related to employment in a definite tar-exposed occupation.

Furthermore, the data showed an excess of bladder cancer in long-exposed workers. These observations are complemented by results of a case-control study carried out by Thériault et al.[25] in the area of the plants. He investigated 81 cases of bladder cancer in the area between 1970 and 1975. From his data he concluded that the relative risk of developing bladder cancer in aluminum workers was 1.9 for nonsmokers and 5.7 for smokers.

Table 2
INCIDENCE OF LUNG CANCER IN NORWEGIAN
ALUMINUM INDUSTRY BY TYPE OF WORK AND
DURATION OF EMPLOYMENT[a]

Department	Length of employment (years)	Standardized incidence ratio		
		Old plants	New plants	All plants
Processing	1.5—4	200 (9)[b]	294 (5)[b]	226 (14)[b]
	5—14	175 (7)	167 (4)	172 (11)[b]
	15 or more	208 (11)[b]	100 (5)	155 (16)
	All	196 (27)[b]	154 (14)	179 (41)[b]
Other	1.5—4	77 (1)	176 (3)	133 (4)
	5—14	91 (1)	100 (2)	97 (3)
	15 or more	133 (4)	128 (5)	130 (9)
	All	111 (6)	132 (10)	123 (16)

[a]   Number of cases in parentheses.
[b] $p < 0.05$ (one-tailed test).

Adapted from Andersen, A., Dahlberg, B. E., Magnus, K., and Wannag, A., *Int. J. Cancer*, 29, 295, 1982.

Andersen et al.[23] studied cancer incidence and mortality in 7410 male employees in four different aluminum plants in Norway during the period 1953 to 1979. For cancer incidence, expected figures were computed on the basis of 5-year, age-specific regional incidence, while national rates were used for mortality. The observed number of cases exceeded that expected for cancer of the lung, kidney, larynx, bladder, and for leukemia. Only for cancer of the lung was the excess statistically significant when compared to the local population. Table 2 shows the data for lung cancer. The excess mortality was more marked in the processing departments than elsewhere, and was more marked in old plants than new plants. The relative risk for workers with more than 25 years of employment in the processing department of older plants was 2.4 times that of the unexposed control groups. On the other hand, in new plants a high was observed: an expected ratio of 2.9 was obtained for process workers with employment for between 1.5 and 4 years. It was suggested that this might be due to an unusually high proportion of cigarette smokers among the short-term workers. Unfortunately, no information on smoking habits was available. A further difficulty in the interpretation of the data results from the great variation of lung cancer incidences in different parts of the country. The calculation of relative risk was therefore crucially dependent on the choice of standard group for calculating the expected number of cases. If it was based on national lung cancer rates, the significance for an excess of lung cancer in aluminum workers vanished.

The largest and probably most comprehensive study of aluminum workers so far has been reported by Rockette and Arena[24] in the U.S. A cohort was formed of 21,829 workers with 5 or more years employment in 14 reduction plants and followed from 1950 to 1977. Three types of processes were used in the plants: prebake (seven plants), vertical-pin Söderberg (one plant), and horizontal-pin Söderberg (five plants). In addition to studying overall mortality patterns for selected causes of death, a more detailed analysis was done for individual processes relative to years of cumulative employment. SMRs were used to compare cause-specific mortality of the workers with that of the U.S. male population. The main results are summarized in Table 3. The results of other studies, which indicate an excess of lung cancer in aluminum workers, were not confirmed. Neither did the duration of employment have a significant influence on cancer mortality. Further examination of the mor-

Table 3
STANDARDIZED MORTALITY RATIO OF U.S.
ALUMINUM WORKERS BY CAUSE OF DEATH AND
TYPE OF PLANT[a]

| | Type of plant | | |
Cause of death	Prebake	Söderberg	All plants[b]
Lung cancer	100 (161)	88 (64)	96 (272)
Other cancer	85 (315)	84 (128)	83 (524)
Major cardiovascular disease	97 (1352)	76 (416)	87 (2093)
Other causes	77 (605)	79 (315)	76 (1065)
All	9∶ (2433)	82 (923)	84(3954)

[a]   Number of deaths in parentheses.
[b]   Including one plant employing both processes.

Adapted from Rockette, M. E. and Arena, V. C., *J. Occup. Med.*, 25, 549, 1983.

tality from types of cancer other than lung cancer showed that there was a slightly raised mortality from cancer of the bladder in the Söderberg plants, and from five other types of cancer when all types of plants were considered together.

C. Coke Oven and Steel Workers

For over 200 years, skin cancers have been recognized to be associated with occupational exposure to tars and pitches related to the destructive distillation of coal. Studies from the early 20th century drew attention to additional cancer sites. Kennaway and Kennaway,[26-28] in a later series of reports, found an increased rate of bladder and lung cancer in occupations involving exposure to coal gas, tar, pitch, and soot.

In an effort to further quantify the potential correlation between occupational exposure and cancer mortality, Doll[29] studied the mortality among male pensioners of a large London gas works company for the period between 1939 and 1948, and compared the data with mortality data for the population of greater London and England and Wales. The pensioners' mortality from cancer was in excess of the expected, and cancer of the lung accounted for the greatest excess (25 against 10.4; $p < 0.001$) which constitutes a significant increase in mortality.

In a separate study, Doll et al.[30] carried out an 8-year prospective analysis of mortality from different causes among several occupational groups of gas workers and retirees covering the years 1953 to 1961. The study included 11,499 men between 40 and 65 years of age at the start of the study with 5 or more years in the gas plant. Observed gas worker mortality rates compared with those expected in populations of England and Wales and regional metropolitan areas.

The workers were grouped into classes according to their exposure: heavy exposure, intermediate exposure, light exposure, or exposure only to by-products. Again, elevated mortality was attributed to respiratory system disease, specifically, cancer of the lung and bronchitis. The lung cancer mortality rate was 69% higher for the heavy exposure group than for the low exposure group. A fourfold higher rate of bladder cancer observed in the heavy exposure group as compared with the light exposure group verged on significance (p = 0.06) according to Doll. He concluded that the mortality of gas workers varied significantly with the type of work and that mortality was highest among workers with greatest exposure to the products of coal carbonization.

A report on an additional 4 years of observation of the cohort[31] provided follow-up information on 2449 coal-carbonizing process workers and 579 maintenance workers for mortality rates gathered at annual intervals from 1961 to 1965. Additional employees of four other gas boards were also followed over periods of 7 to 8 years. Cause of death in the original studies was obtained from death certificates. The new data showed a pattern similar to that of data for the first 8 years.

Heavily exposed workers experienced a highly significant elevated mortality from lung cancer ($p < 0.001$) and bronchitis ($p < 0.001$). Data on by-product workers showed no excessive mortality, and over the 12-year period provided no substantial evidence of increased occupational risk of this group. The additional 4 years of data in this study support the earlier association between exposure to the products of coal carbonization and increased lung cancer and also a risk of bladder cancer ($p = 0.06$).

Recently, Manz et al.[32] reported on a cohort study of workers employed in gasworks in Hamburg. A total of 4516 male workers, who had been employed for 10 years or more, were followed up since 1900. All but 52 men were traced. Six exposure groups were differentiated and were assembled into three cohorts: (1) retort house workers, (2) other technical personnel, and (3) office and administrative personnel.

When comparing these cohorts with each other, it was seen that, for retort house workers, generally there existed a high risk of developing malignant tumors of the respiratory tract as well as the urinary tract. When using expected values from cohort 2, the relative risk of cancer of lung for retort house workers was 3.53, and 8.24 when compared to cohort 3. The relative risk of bladder cancer for retort house workers was determined to be 4.35 compared to pooled cohorts 2 and 3. The significant increases in lung and bladder cancer mortality in retort house workers were attributed to the exposure of coal tar pitch volatiles.

Several studies of mortality among coke oven workers in the iron and steel industry have confirmed and extended the established findings that workers involved with coal-carbonization processes experience a markedly increased cancer risk.[21]

In the first phase of a long-term study of workers in the U.S. steel industry, Lloyd[33] undertook a 9-year prospective analysis of 2552 coke plant workers employed in 1953. He examined the mortality records of the workers in relation to length of employment and work area within the coke plant, and compared the cause-specific mortality of coke plant workers as a whole with the mortality of the total steelworker population of 58,828 workers. Coke plant workers were categorized as oven workers and nonoven workers. The excess lung cancer mortality among coke plant workers indicated that risk was elevated nearly threefold among coke oven workers (20 observed deaths vs. 7.5 expected). Men working on the tops of the coke ovens had nearly a fivefold increased risk, and men employed 5 or more years at full-time topside jobs had a tenfold risk (15 observed deaths vs. 1.5 expected). Furthermore, a significant excess of cancer of the digestive system was observed in both long- and short-term nonoven workers (17 vs. 9.7 expected, significant at the 5% level). Cancer of the pancreas and large intestine accounted for the greatest excess.

Redmond et al.,[34] in a follow-up of earlier reports in the series, examined the mortality records of cohorts of coke oven workers in an expanded study at 12 steel plants. In addition, the data from the earlier study[33] of two Allegheny County steel plants with coke plants were updated to 1966 and were compared with data from 10 other plants for the same period. The cohorts at the ten additional plants included all men who had worked at or on the oven at any time in the 5-year period, 1951 through 1955. (The criterion for inclusion in the prior Allegheny County study was employment in one of seven steel plants during 1953).

The findings of Redmond et al.[34] further indicate that both the level and duration of exposure to coke oven emissions are correlated with mortality from various types of cancer, further demonstrating the additivity of time and concentration.

Overall mortality of white coke oven workers is somewhat less than expected. When mortality is categorized by cause, the rates for coke oven workers are significantly elevated

Table 4
RELATIVE RISK OF DEATH FROM CANCER OF
THE RESPIRATORY SYSTEM FOR COKE OVEN
WORKERS, 1953—1970[a]

Work area	Employed 5+ years	Employed 10+ years	Employed 15+ years
Coke oven	3.02 (54)	3.42 (44)[b]	4.14 (33)[b]
Oven topside full-time	9.19 (25[b]	11.79 (16)[b]	15.72 (8)[b]
Oven topside part-time	2.29 (12)[b]	3.07 (16)[b]	4.72 (18)[b]
Oven side only	1.79 (17)[c]	1.99 (12)[b]	2.00 (7)[b]

[a]   Number of observed deaths given in parentheses.
[b] $p < 0.01$.
[c] $p < 0.05$.

From Redmond, C. K., *Environ. Health Perspect.*, 52, 67, 1983. With permission.

for malignant neoplasms (relative risk, RR 1.34; $p < 0.01$), for malignant neoplasms associated primarily with respiratory cancer (RR 2.85; $p < 0.01$), for kidney cancer (RR 7.49; $p < 0.01$), and for prostate cancer (RR 1.64; not significant).

The study showed that men employed at full-time topside jobs for 5 years or more have relative risk of lung cancer of 6.87 ($p < 0.01$) compared to RR 3.22 ($p < 0.01$) for men with 5 years of mixed topside and side-oven experience and RR 2.10 ($p < 0.05$) for men with more than 5 years of side-oven experience. These data indicate a definite gradient in response based on both type and duration of exposure.

Overall, the study confirmed the Lloyd[33] findings of a more than twofold excess of mortality due to respiratory cancers in all coke oven workers. A new finding of Redmond's was a significant excess of kidney cancer among coke oven workers (RR 7.49; $p < 0.01$).

Redmond concludes that the RR 6.87 ($p < 0.01$) for malignant neoplasms of the respiratory system for men employed full-time topside, and the RR 1.70 (not significant) for men employed less than 5 years constituted further evidence of a dose-response relationship. In later follow-up studies, Redmond[35-36] confirmed elevated risks for coke oven workers for lung cancer.

Table 4 summarizes the findings noted for respiratory cancer among coke oven workers between 1953 and 1970. A strong relationship is observed for increased risks associated with longer duration of exposure and intensity of exposure. Among the high-exposed topside workers with 15 years or more experience, the relative risk is almost 16-fold. For oven side workers no or only slight increase in lung cancer mortality is observed.

Significant risks of cancer of the genito-urinary organs, especially kidney, were also observed for coke oven workers. For nonoven workers, an excess of deaths from cancer of the large intestine and pancreas was observed.[36]

In 1980, Lloyd[37] presented results from the updated study of the coke oven population of his earlier study[33] through 1975. With 30.9% of the original population from Allegheny County plants deceased, the relative risk for lung cancer in men employed for 5 years or more in 1953 and working full-time topside was 6.94. This was significant and based on 25 deaths instead of 15 in 1961. Lloyd also pointed out the possible underestimation of relative risk that could result from the use of internal controls, since a high risk of lung cancer has been demonstrated in many other areas in steel mills compared to rates in the general population.

Although in steel plants the greatest risk of cancer of the lung is at the top of the coke ovens, recent observations indicate significant excess of death due to lung cancer also in

Table 5

**OBSERVED AND EXPECTED DEATHS AND STANDARDIZED
MORTALITY RATIOS (SMR) BY RACE FOR THOSE WORK AREAS
SHOWING A SIGNIFICANTLY HIGH SMR FOR CANCER OF LUNGS,
BRONCHUS, AND TRACHEA**

Work area	White			Nonwhite		
	Observed	Expected	SMR	Observed	Expected	SMR
Coke plant	27	22.3	121	56	13.6	410[a]
Open heart	110	75.5	146[b]	44	27.2	162[b]
Foundry	28	14.0	201[b]	2	4.1	—
Billet, bloom, and slab	67	46.6	144[b]	4	4.0	—
Structural mills	27	17.6	154[b]	3	3.2	—
Cold reducing mill	33	16.4	201[b]	0	1.5	—
Hot pack mills	20	11.3	177[b]	0	0.1	—
Hot strip rolling	28	15.5	181[b]	0	0.0	—
General labor	12	6.9	173	14	6.5	216[b]
Mason department	23	12.6	182[b]	10	5.2	191
Blacksmiths shop	11	4.3	256[b]	0	0.0	—
Elec. main. assq.	40	24.8	161[b]	0	0.0	—
Machine shop	34	20.9	163[b]	0	0.0	—
Paint shop	9	4.1	222[b]	0	0.0	—
Riggers and structural	35	18.6	188[b]	0	0.0	—
All other	6	1.9	320[b]	0	0.0	—

[a]   Significance, $p < 0.01$.
[b]   Significance, $p < 0.05$.

From Lloyd, J. W., *Luftverunreinigung durch Polycyclische Kohlenwasserstoffe*, VDI-Berichte Nr. 358,
VDI-Verlag, Düsseldorf, 1980, 237. With permission.

other work areas. Table 5 shows more work areas in Allegheny County steel plants showing
a significant excess in lung cancer in comparison to the U.S. population. As previously
found, the highest mortality ratio is for nonwhite coke plant workers. In addition, excesses
of lung cancers are observed for workers in a great number of other work areas. Lloyd[37]
suggested for further study the foundry, the open hearth, and the general labor group because
of the following reasons: there have been several other studies indicating a possible lung
cancer problem in foundries, and the open hearth and the general labor group show excess
for the two race groupings.

D. Iron and Steel Foundry Workers

Pyrolysis products of organic materials used in mold binders and additives are common
in foundry emissions. These pyrolysis products include PAH which have been detected in
many locations in foundries.[38,39]

The epidemiology of lung cancer in foundries has been reviewed by Tola,[40] Palmer and
Scott,[41] and IARC.[21]

The first reports of excess lung cancer mortality among foundry workers came from
England in 1938 and were based on official death records and census figures.[42] Later, several
other surveys based on mortality statistics indicated also that foundry work may be associated
with an increased risk of lung cancer.[21] Frequently, however, biases exist in studies based
on nonliving populations, and surveys of mortality statistics are known to suffer from many
limitations.

By far the most useful technique to detect an occupational hazard is the cohort study, in
which a defined group of workers is followed over a set period of time to discover how

many have died from the specified causes of death. The first lung cancer mortality study in foundries to employ the retrospective cohort mortality method was performed in Finland.[43] The cohort consisted of 3876 men employed in 20 representative foundries between 1950 and 1973. Of the three categories of foundries studied, elevated lung cancer mortality rates were observed only in iron foundries. The highest risk was observed for those employed for a minimum of 5 years (SMR = 270).

In contrast to the Finnish study, a retrospective cohort study in a Canadian steel mill indicated an increased risk of lung cancer for steel foundry workers with 5 or more years of exposure.[44] SMRs were based on mortality rates of a nearby metropolitan area. The overall SMR was 250 for lung cancer among foundrymen. Crane operators had the highest risk for lung cancer (SMR = 714) followed by fettlers (SMR = 314), molders (SMR = 225), and coremakers (SMR = 208).

Decouflé and Wood[45] studied the mortality patterns among workers employed between 1938 and 1967 in a captive gray iron foundry located in a northern U.S. city. The observed mortality rates of the cohort, which consisted of 2861 foundrymen, were compared with those of the general U.S. population. The cohort was subdivided into three groups according to length of employment. The only statistically significant increase in the ratio of observed to expected numbers of deaths from cancer was found in the group that had been employed for 5 or more years prior to 1938. A twofold excess in mortality was observed for both cancer of the digestive system and respiratory cancer.

On the other hand, an extensive study on steelworkers in seven U.S. steel plants did not reveal a significant increase of lung cancer mortality among foundry workers compared to that of the total steelworker population under study.[46,47] However, a significant excess of deaths from cancer of the genito-urinary system was observed.

Recently, Fletcher and Ades[48] studied the mortality experience of workers who had started work between 1946 and 1965 in English steel foundries. The cohort was followed prospectively until 1978. Expected deaths were calculated on the basis of death rates in England and Wales. Mortality from lung cancer was significantly increased among workers employed in the foundry and fettling shop areas (SMR = 142 and 173, respectively). There was some evidence of risk increasing with length of employment.

In most of the epidemiological studies cited previously, the study populations were relatively small, and in some cases, the number of cancer cases available for study were too few to meet the statistical requirements. While no one of these reports may be conclusive alone, the evidence for the association between an excess risk of lung cancer and foundry work is sufficiently consistent that it cannot be ignored. PAH, which are known to occur in foundry atmospheres, are suspected among the etiological factors.

E. Asphalt and Tar Workers

Results of a health survey of 462 asphalt workers from 7 asphalt companies and a control group from 25 oil refineries were published in 1968 by Baylor and Weaver.[49] The physical examination emphasized the skin and the respiratory tract. One case of lung cancer had previously been diagnosed in one of the workers in the control group. Skin cancer was present in four asphalt workers, but two of these workers had skin cancer prior to beginning work in the asphalt operations.

Hammond et al.[50] conducted a prospective study of mortality from 1960 through 1971 among 5939 male hot pitch roofers and waterproofers. Motivation for the study was provided by the knowledge that most of these workers were frequently exposed to extremely high concentrations of BaP in the air they breathed. The principal criterion for inclusion of an individual in the cohort was that he must have been a member of the United Slate, Tile and Composition Roofers, Damp and Waterproof Workers' Association for at least 9 years prior to the onset of follow-up on January 1, 1960. The inclusion of workers irrespective of whether active, probational, or in retired status tended to reduce the healthy-worker bias which would apply to the follow-up of only active workers. The quality of follow-up for mortality was good: only 151 men were unsuccessfully traced with respect to survival status during the 12-year follow-up period; 1798 deaths were recorded in all.

The cohort person-years of experience were classified according to years of past membership in the union. Observed deaths in major cause categories were compared with expected numbers. The latter were obtained by applying age- and calendar year-specific rates for the entire male population of the U.S. to the corresponding person-years of exposure in the cohort classes.

In all age classes except 50 to 59 years, the ratios of observed to expected deaths slightly exceeded 1. Thus, if there were a healthy-worker bias, it was more than exceeded by generally higher rates of deaths in the worker cohort than among the general population of U.S. males. The overall mortality ratio for workers with only 9 to 19 attained years since joining the union was 1.02, compared to 1.09 for those with 20 or more attained years since joining. The respective cancer mortality ratios of 1.07 and 1.45 showed the levels of these ratios were not generally greater than for certain other causes of death such as accidents (1.59 and 1.41, respectively) and deaths from other respiratory causes (1.96 and 1.67, respectively). The mortality ratios for all other causes than specified above, however, were less than 1. For cancer of the lung, the mortality ratio rose from 0.92 for 9 to 19 years of membership to 2.47 for 40 or more years. Taken in itself, the latter gradient might be dismissed as merely the result of an attenuation in the healthy-worker bias with the passage of time. However, similar gradients were not seen for the mortality ratios in the other major cause of death categories. Thus, there was support for the concept that there was an increased hazard of lung cancer with extended duration of exposure.

III. SUMMARY AND CONCLUSION

Studies of occupational hazards related to coal tar pitch volatiles in general and PAH in particular have a long history. The association of these compounds with the development of cancer has been recognized for more than 200 years. Chemical analytical studies show that PAH appear in a large number of industrial processes, mainly due to high temperature treatment of coal tar and pitch as well as incomplete combustion or pyrolysis of organic material in general. Furthermore, biological and toxicological studies show that many PAH compounds exhibit carcinogenic effects in experimental animals, demonstrating that PAH is the largest class of chemical carcinogens known today.

Although no work environment has PAH as its only pollutant, epidemiological studies of workers exposed mainly to PAH indicate an increased risk of lung cancer related to PAH exposure in work atmospheres. For a single occupation there seems to be a correlation between duration and degree of exposure (tar-years) and cancer incidence. However, comparing levels of exposure in different occupations and industries, there is no apparent correlation with cancer incidences. Based on the animal carcinogenic data and the epidemiological evidence, PAH must be regarded as an occupational hazard to which exposure should be as low as possible.

REFERENCES

1. **Pott, P.,** Cancer scroti, in *Chirurgical Observations,* 1775; reprinted in *Natl. Cancer Inst. Monogr.,* 10, 7, 1963.
2. **Volkmann, R. von,** *Beiträge zur Chirurgie,* Breitkopf und Härtel, Leipzig, 1875.

3. **Bell, J.**, Paraffin epithelioma of the scrotum, *Edinburgh Med. J.*, 22, 153, 1876.

4. **Butlin, H. T.**, Three lectures on cancer of the scrotum in chimney sweepers and others, *Br. Med. J.*, 2, 1341; 3, 1; 3, 66, 1892.

5. **Yamagiwa, K. and Ichikawa, K.**, Über die künstliche Erzeugung von Papillom, *Verh. Jpn. Pathol. Ges.*, 5, 142, 1915.

6. **Tsutsui, H.**, Über das künstlich erzeugte Canceroid bei der Maus, *Gann*, 12, 17, 1918.

7. **Kekulé, A.**, Untersuchungen über aromatische Verbindungen, *Liebigs Ann. Chem. Pharmacol.*, 137, 129, 1865.

8. **Berthelot, M.**, Les polymères de l'acetylène. I. Synthèse de la benzine, *C. R. Acad. Sci.*, 63, 479, 1866.

9. **Hieger, I.**, Spectra of cancer-producing tars and oils and of related substances, *Biochem. J.*, 24, 505, 1930.

10. **Kennaway, E. L. and Hieger, I.**, Carcinogenic substances and their fluorescence spectra, *Br. Med. J.*, 1044, 1930.

11. **Hieger, I.**, The isolation of a cancer-producing hydrocarbon from coal tar, *J. Chem. Soc.*, 395, 1933.

12. Committee on Biologic Effects of Atmospheric Pollutants, *Particulate Polycyclic Organic Matter*, National Academy of Sciences, Washington, D.C., 1972.

13. **Dipple, A.**, Polynuclear aromatic carcinogens, in *Chemical Carcinogens*, Searle, C. E., Ed., ACS Monogr. 173, American Chemical Society, Washington, D. C., 1976, chap. 5.

14. International Agency for Research on Cancer, *IARC Monographs on the Evaluation of the Carcinogenic Risk of Chemicals to Humans*, Polynuclear Aromatic Compounds, Part 1, Chemical, Environmental and Experimental Data, IARC, Lyon, France, 1983.

15. **Henry, S. A.**, The study of fatal cases of cancer of the scrotum from 1911 to 1935 in relation to occupation, with special references to chimney sweeping and cotton mule spinning, *Am. J. Cancer*, 31, 28, 1937.

16. **Kupetz, G. W.**, Krebs des Atemtraktes und Schornsteinfegerberuf, Dissertation, Humboldt-Universität, Berlin, 1966.

17. **Hansen, E. S.**, Mortality of Danish chimney sweeps 1970-1975, in *Int. Symp. Prevention of Occupational Cancer*, Institute of Occupational Health, Helsinki, 1981, 104.

18. **Hogstedt, C., Anderson, K., Frenning, B., and Gustavsson, A.**, A cohort study on mortality among long-time employed Swedish chimney sweeps, *Scand. J. Work Environ. Health*, 8 (Suppl. 1), 72, 1982.

19. **Kreyberg, L.**, 3:4 Benzpyrene in industrial air pollution: some reflexions, *Br. J. Cancer*, 13, 618, 1959.

20. **Konstantinov, V. G. and Kuz'minykh, A. I.**, Resinous substances and 3,4-benzpyrene in the air of electrolysis shops of aluminum plants and their role in carcinogenesis (in Russian), *Gig. Sanit.*, 36, 39, 1971; *Chem. Abstr.*, 74, 145994g, 1971.

21. International Agency for Research on Cancer, *IARC Monographs on the Evaluation of the Carcinogenic Risk of Chemicals to Humans*, Polynuclear Aromatic Compounds, Part 3, Industrial Exposure in Aluminium Production, Coal Gasification, Coke Production, and Iron and Steel Founding, IARC, Lyon, France, 1984.

22. **Gibbs, G. W. and Horowitz, I.**, Lung cancer mortality in aluminum reduction plant workers, *J. Occup. Med.*, 21, 347, 1979.

23. **Andersen, Aa., Dahlberg, B. E., Magnus, K., and Wannag, A.**, Risk of cancer in the Norwegian aluminum industry, *Int. J. Cancer*, 29, 295, 1982.

24. **Rockette, M. E. and Arena, V. C.**, Mortality studies of aluminum reduction plant workers: potroom and carbon department, *J. Occup. Med.*, 25, 549, 1983.

25. **Thériault, G., DeGuire, L., and Corchier, S.**, Reducing aluminum: an occupation possibly associated with bladder cancer, *Can. Med. Assoc. J.*, 124, 1981.

26. **Henry, S. A., Kennaway, N. M., and Kennaway, E. L.**, The incidence of cancer of the bladder and prostate in certain occupations, *J. Hyg.*, 31, 125, 1931.

27. **Kennaway, N. M. and Kennaway, E. L.**, A study of the incidence of cancer of the lung and larynx, *J. Hyg.*, 36, 236, 1936.

28. **Kennaway, E. L. and Kennaway, N. M.**, A further study of the incidence of cancer of the lung and larynx, *Br. J. Cancer*, 1, 260, 1947.

29. **Doll, R.**, The causes of death among gas workers with special reference to cancer of the lung, *Br. J. Med.*, 9, 180, 1952.

30. **Doll, R., Fisher, R. E. W., Gammon, E. J., Gunn, W., Hughes, G. O., Tyrer, F. H., and Wilson, W.**, Mortality of gas workers with special reference to cancer of the lung and bladder, chronic bronchitis, and pneumonoconiosis, *Br. J. Ind. Med.*, 22, 1, 1965.

31. **Doll, R., Vessey, M. P., Beasley, R. W. R., Buckley, A. R., Fear, E. C., Fisher, R. E. W., Gammon, E. J., Gunn, W., Hughes, G. O., Lee, K., and Norman-Smith, B.**, Mortality of gasworkers — final report of a prospective study, *Br. J. Ind. Med.*, 29, 394, 1972.

32. **Manz, A., Berger, J., and Waltsgott, H.**, Zur Frage des Berufskrebses bei Beschäftigten der Gasindustrie, *Forschungsbericht Nr. 352*, Bundesanstalt für Arbeitsschutz und Unfallforschung, Dortmund, 1983.

33. **Lloyd, J. W.**, Longterm mortality study of steelworkers. V. Respiratory cancer in coke plant workers, *J. Occup. Med.*, 13, 52, 1971.

34. **Redmond, C. K., Ciocco, A., Lloyd, J. W., and Rush, H. W.,** Longterm mortality study of steelworkers. VI. Mortality from malignant neoplasma among coke oven workers, *J. Occup. Med.*, 14, 621, 1972.
35. **Redmond, C. K., Strobino, B. R., and Cypress, R. H.,** Cancer experience among coke by-product workers, *Ann. N. Y. Acad. Sci.*, 217, 102, 1976.
36. **Redmond, C. K.,** Cancer mortality among coke oven workers, *Environ. Health Perspect.*, 52, 67, 1983.
37. **Lloyd, J. W.,** Problems of lung cancer mortality in steelworkers, in *Luftverunreinigung durch Polycyclische Kohlenwasserstoffe*, VDI-Berichte Nr. 358, VDI-Verlag, Düsseldorf, 1980, 237.
38. **Schimberg, R. W., Skyttä, E., and Falck, K.,** Iron foundry workers' exposure to polycyclic aromatic hydrocarbons (in German), *Staub-Reinhalt. Luft*, 41, 421, 1981.
39. **Verma, D. K., Muir, D. C. F., Cunliffe, S., Julian, J. A., Vogt, J. H., and Rosenfeld, J.,** Polycyclic aromatic hydrocarbons in Ontario foundry environments, *Am. Occup. Hyg.*, 25, 17, 1982.
40. **Tola, S.,** Epidemiology of lung cancer in foundries, *J. Toxicol. Environ. Health*, 6, 275, 1980.
41. **Palmer, W. G. and Scott, W. D.,** Lung cancer in ferrous foundry workers: a review, *Am. Ind. Hyg. Assoc. J.*, 42, 329, 1981.
42. **Turner, H. M. and Grace, H. G.,** An investigation into cancer mortality among males in certain Sheffield trades, *J. Hyg.*, 38, 90, 1938.
43. **Koshela, R.-S., Hernberg, S., Kärävä, R., Järvinen, E., and Nurminen, M.,** A mortality study of foundry workers, *Scand. J. Work. Environ. Health*, 2, (Suppl. 1), 73, 1976.
44. **Gibson, E. S., Martin, R. H., and Lockington, J. N.,** Lung cancer mortality in a steel foundry, *J. Occup. Med.*, 19, 807, 1977.
45. **Decouflé, P. and Wood, D. J.,** Mortality patterns among workers in a gray iron foundry, *Am. J. Epidemiol.*, 109, 667, 1979.
46. **Lloyd, J. W., Lundin, F. E., Jr., Redmond, C. K., and Geiser, P. B.,** Long-term mortality by work area, *J. Occup. Med*, 12, 151, 1970.
47. **Redmond, C. K., Wieand, H. S., Rockette, H. E., Sass, R., and Weinberg, G.,** Long-term Mortality Experience of Steelworkers, Publ. No. 81—120, National Institute of Occupational Safety and Health. U.S. Department of Health and Human Services, Cincinnati, 1981.
48. **Fletcher, A. C. and Ades, A.,** Lung cancer mortality in a cohort of English foundry workers, *Scand. J. Work Environ. Health*, 10, 7, 1984.
49. **Baylor, C. H. and Weaver, N. K.,** A health survey of petroleum asphalt workers, *Arch. Environ. Health*, 17, 210, 1968.
50. **Hammond, E. C., Selikoff, I. J., Lawther, P. L., and Seidman, H.,** Inhalation of benzo(a)pyrene and cancer in man, *Ann. N. Y. Acad. Sci.*, 271, 116, 1976.

Chapter 2

MODE OF PAH FORMATION

PAH can be formed by thermal decomposition of any organic materials containing carbon and hydrogen. The formation is based on two major mechanisms: (1) pyrolysis or incomplete combustion, and (2) carbonization processes. In addition, some publications suggest a biological formation of PAH. This chapter will briefly review these mechanisms.

Although the mechanism of PAH formation in combustion processes is complex and variable, a pioneering contribution to the understanding has been given by Badger and co-workers.[1,2] The chemical reactions in flames proceed by free radical paths, and a synthetic route based on this concept is postulated for the formation of PAH also. Based upon the results of a series of pyrolysis experiments, Badger suggested the stepwise synthesis of PAH from C_2 species during hydrocarbon pyrolysis as outlined in Figure 1 for benzo(a)pyrene (BaP) as a example. These pyrolysis studies were conducted by passing the hydrocarbon vapor in nitrogen through a silica tube at 700°C. Although the use of nitrogen atmospheres has been criticized as lacking relevance to actual combustion, the reducing conditions are similar to those of the oxygen-deficient environments common in the center of flames and the data are in good qualitative agreement with the PAH combustion products formed. For example, Boubel and Ripperton[3] found that BaP is produced during combustion even at high percentages of excess air, although the amount of BaP is larger at lower percentages of excess air.

Lending support to the postulated route to PAH, Badger and Spotswood[4] pyrolyzed toluene, ethylbenzene, propylbenzene, and butylbenzene and obtained the highest yields of BaP with butylbenzene, a potential intermediate in Badger's reaction scheme. Obviously it is unnecessary to break the starting material down completely to a two-carbon radical in order to form BaP. Any component of the combustion reaction that can contribute intermediate pyrolysis products of the structure required for BaP synthesis would be expected also to lead to increased yields of BaP. It was also found that, when 1,3-butadiene was pyrolyzed with pyrene, no increase in the yield of BaP was observed, indicating that Diels-Alder type reactions are probably not important.

More recent studies tend to confirm most of the mechanism proposed by Badger. Crittenden and Long[5] determined the chemical species formed in rich oxy-acetylene and oxy-ethylene flames. Compounds identified suggest that the C_2 species react to form C_4, C_6, and C_8 species, and that reactions involving styrene and phenylacetylene are probably important in the formation of PAH. Also, a $C_{10}H_{10}$ species was detected in the gases of both flames, which corresponds to the C_4 substituted benzene postulated by Badger (Structure 4 in Figure 1).

The mechanism in Figure 1 is a possible pathway to BaP formation, but similar routes could be devised with different intermediates, to lead to most of the known PAH produced in combustion processes. Badger's work, with its reliance on calculated C-C and C-H bond energies to predict favored pathways and the experimental confirmation of these steps with radioisotopic labeling, provides a plausible mechanism for PAH formation in combustion or pyrolytic processes.

Once formed, PAH might undergo further pyrolytic reactions to form larger PAH by intermolecular condensation and cyclization. This has been extensively studied by Zander and co-workers.[6] With nonsubstituted PAH, polyaryls are formed, to be followed by ring closure to highly condensed hydrocarbons. Figure 2 shows the pyrolytic formation of 5-ring PAH from naphthalene as an example. The three possible biaryls 2 to 4 are converted to the pericondensed systems 5 to 8. Cyclization products from 4 are benzologs of the relatively unstable biphenylene and are not observed.

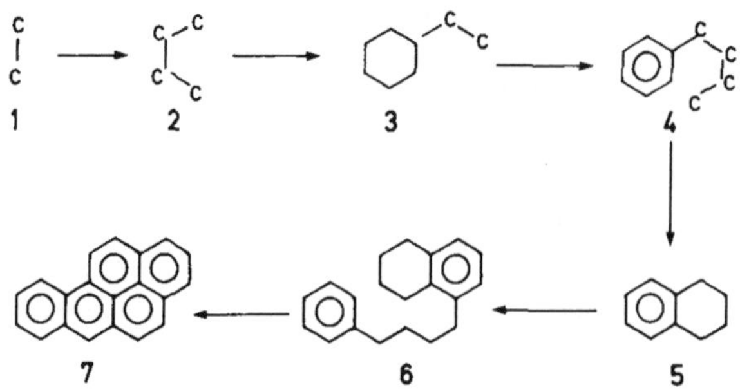

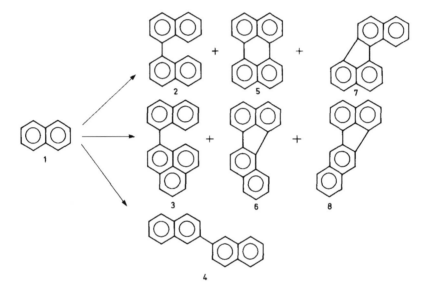

FIGURE 1.   Pyrolytic formation of benzo(a)pyrene after Badger.[1]

FIGURE 2.   Pyrolytic formation of larger PAH from smaller ones.[6]

In spite of the tremendous number of different PAH which might be formed during the primary reactions, only a limited number of PAH enters the environment. Many of the primarily formed PAH will have short half-lives under the pyrolysis conditions and will stabilize in the following reactions. At high temperatures the thermodynamically most stable compounds will be formed in corresponding quantitative ratios. These are mainly the unsubstituted parent PAH. Irrespective of the type of material to be burned, surprisingly similar ratios of PAH are formed at a defined temperature. For example, thermal decomposition of pit coal, cellulose, tobacco, and also of polyethylene and polyvinylchloride which is carried out at 1000°C yields very similar PAH profiles.[7] Consequently, PAH profiles seem to depend more on the combustion conditions rather than on the type of organic material burned.

The absolute amount of PAH formed under defined pyrolysis conditions depends, however, on the temperature as well as the material. The amount of BaP formed on pyrolysis of various substances at 840°C in a nitrogen stream is shown in Table 1.[8] The table demonstrates the relative unimportance of the oxygen-containing carbohydrates in the generation of BaP as compared to the C_{32} paraffin or β-Sitosterol, a common plant sterol. Furthermore, the absolute amount of PAH formed during incomplete combustion is dependent on temperature, as shown in Table 2. Under comparable conditions, 1 g of tobacco yields 44 ng of BaP at 400°C and 183,500 ng of BaP at 1000°C.[9]

At temperatures below approximately 700°C the pyrolysis products contain, besides the

Table 1
LEVELS OF BENZO(A)PYRENE
(BaP) PRODUCED ON PYROLYSIS
(840°C,N$_2$)

Substance pyrolyzed	μg BaP/g pyrolyzed
Glucose	47.5
Fructose	98.4
Cellulose	288.8
Stearic acid	1200
Dotriacontane	3130
β-Sitosterol	3750

From Schmeltz, I. and Hoffmann, D., *Carcinogenesis — A Comprehensive Survey*, Vol. 1, Freudenthal, R. I. and Jones, P. W., Eds., Raven Press, New York, 1976, 225. With permission.

Table 2
FORMATION OF BENZO(A)PYRENE FROM BLEND
TOBACCO (100 g) AS A FUNCTION OF TEMPERATURE
(1000 mℓ N$_2$/min)

Temperature (°C)	Condensate (g)	Nitromethane extract (g)	BaP (μg)	BeP (μg)	e/a
400	18.6850	1.8820	4.4	3.2	0.73
500	18.8966	1.9260	12.9	8.4	0.65
600	15.2240	1.7201	32.0	19.0	0.60
700	11.2810	1.5622	56.0	29.4	0.53
700	10.3490	1.4931	88.6	45.2	0.51
800	7.8900	1.2725	270.0	155.0	0.57
900	4.9921	1.2441	1820.0	824.0	0.45
900[a]	5.0120	1.3330	4725.0	2015.0	0.43
1000[a]	4.6941	1.2320	18350.0	6710.0	0.36

[a] At 1500 mℓ/min to avoid a back reaction of the condensate.

From Grimmer, G., *Environmental Carcinogens: Polycyclic Aromatic Hydrocarbons*, CRC Press, Boca Raton, Fla., 1983. With permission.

parent PAH, also larger amounts of alkyl-substituted PAH, mainly methyl derivatives. A typical example is tobacco smoke which yields soot quite abundantly in alkyl-substituted PAH.[10] Figure 3 shows schematically the relative abundance of PAH as a function of the number of alkyl carbons at different formation temperatures as given by Blumer.[11] As indicated by the figure, the number of alkyl carbons present as side chains on PAH correlates closely with the temperature at which the compounds are formed.

PAH is not formed solely by high temperatures and open flames. Various processes of carbonization that occur, e.g., during the generation of mineral oil and coal, lead to the formation of PAH from decaying biological material at low temperatures (less than 200°C) over a period of millions of years. The transformation resembles pyrolysis, but the reactions are exceedingly slow because of the modest temperatures involved.

The profile of PAH contained in fossil fuels is clearly different from that obtained by incomplete combustion or pyrolysis of organic material (see Figure 3). As the formation temperature of fossil fuels is quite low, large amounts of alkyl-substituted PAH are present

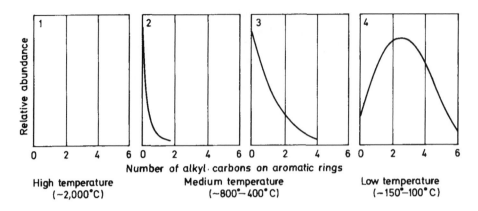

FIGURE 3.   Relative abundance of alkyl carbons on PAH rings. (From Blumer, M., *Sci. Am.*, 234, 34, 1976. With permission.)

in crude oil and coal-derived materials. Furthermore, mineral oil frequently contains predominantly one of several possible isomeric forms of PAH. This applies, for example, to the two isomeric benzopyrenes. In mineral oil, benzo(e)pyrene is the predominant form with much less BaP present.[12] In exhaust gas condensate from gasoline or diesel engines, the ratio of the two isomeric benzopyrenes is about 1:1.[7] This fact reflects the much greater time available for petroleum formation to accomplish energetically favored reaction pathways.

The biosyntheses of PAH by microorganisms, algae, and plants have been reported.[13] However, criticism was made of the experimental work, especially the way in which blank measurements were conducted. Grimmer could show that when strict care was taken to exclude all outside contamination, no PAH could be found in plants.[11] In any case, the amounts of PAH possibly formed in biosynthesis will hardly be of importance in comparison to the quantities formed abiotically.

SUMMARY AND CONCLUSION

PAH are formed by thermal decomposition of any organic material containing carbon and hydrogen. During incomplete combustion or pyrolysis, PAH are supposed to be generated from small fragments via radical pathways. Once formed, PAH may undergo further pyrolytic reactions to larger PAH by intermolecular condensation and cyclization. The amount of PAH formed depends on both the type of organic material and the temperature. The relative distribution of PAH components depends mainly on the temperature.

PAH in fossil fuels have been formed at low temperatures by a carbonization process over millions of years. The PAH profile contained in fossil fuels is clearly different from that obtained by pyrolysis. In petroleum, the alkylated PAH outnumber the parent PAH.

The amount of PAH possibly formed by biosynthesis in microorganisms or plants is negligible in comparison with the quantities formed by abiotic processes.

REFERENCES

1. **Badger, G. M.,** Mode of formation of carcinogens in human environment, *Natl. Cancer Inst. Monogr.*, 9, 1, 1962.
2. **Badger, G. M., Ed.,** *The Chemical Basis of Carcinogenic Activity*, Charles C Thomas, Springfield, Ill., 1962.

3. **Boubel, R. W. and Ripperton, L. A.**, Benzo(a)pyrene production during controlled combustion, *J. Air Pollut. Control Assoc.*, 13, 553, 1963.

4. **Badger, G. M. and Spotswood, T. M.**, The formation of aromatic hydrocarbons at high temperatures. IX. The pyrolysis of toluene, ethylbenzene, propylbenzene and butylbenzene, *J. Chem. Soc.*, 4420, 1960.

5. **Crittenden, B. D. and Long, R.**, The mechanisms of formation of poly-nuclear aromatic compounds in combustion systems, in *Carcinogenesis — A Comprehensive Survey*, Vol. 1, Freudenthal, R. I. and Jones, P. W., Eds., Raven Press, New York, 1976, 209.

6. **Lang, K. F., Buffleb, H., and Zander, M.**, Pyrolysen von mehrkernigen aromatischen Kohlenwasserstoffen, *Erdöl Kohle*, 16, 944, 1963.

7. **Grimmer, G., Ed.**, *Environmental Carcinogens: Polycyclic Aromatic Hydrocarbons*, CRC Press, Boca Raton, Fla., 1983.

8. **Schmeltz, I. and Hoffman, D.**, Formation of polynuclear aromatic hydrocarbons from combustion of organic matter, in *Carcinogenesis — A Comprehensive Survey*, Vol. 1, Freudenthal, R. I. and Jones, P. W., Eds., Raven Press, New York, 1976, 225.

9. **Grimmer, G., Glaser, A., and Wilhelm, G.**, Die Bildung von Benzo(a)pyren und Benzo(e)pyren beim Erhitzen von Tabak in Abhängigkeit von Temperatur und Strömungsgeschwindigkeit in Luft- und Stickstoffatmosphäre, *Beitr. Tabakforsch.*, 3, 415, 1966.

10. **Severson, R. F., Schlotzhauer, W. S., Arrendale, R. F., Snook, M. E., and Higman, H. C.**, Correlation of polynuclear aromatic hydrocarbon formation between pyrolysis and smoking, *Beitr. Tabakforsch.*, 9, 23, 1977.

11. **Grimmer, G.** as cited in **Blumer, M.**, Polycyclic aromatic compounds in nature, *Sci. Am.*, 234, 34, 1976.

12. **Guerin, M. R.**, Energy sources of polycyclic aromatic hydrocarbons, in *Polycyclic Hydrocarbons and Cancer*, Vol. 1, Gelboin, H. V. and Ts'o, P. O. P., Eds., Academic Press, New York, 1970, chap.1.

13. **Suess, M. J.**, The environmental load and cycle of polycyclic aromatic hydrocarbons, *Sci. Total Environ.*, 6, 239, 1976.

Chapter 3

PHYSICAL AND CHEMICAL PROPERTIES OF PAH

I. CHEMICAL STRUCTURE AND NOMENCLATURE

PAH consist of three or more benzene rings interlinked in various arrangements. The entire group of PAH can be divided into kata-annellated and peri-condensed systems. In kata-annellated PAH, the tertiary carbon atoms are centers of two interlinked rings, e.g., anthracene (1), whereas in peri-condensed PAH, some of the tertiary C atoms are centers of three interlinked rings, e.g., pyrene (2). Annellation can be linear (anthracene,1) or angular (phenanthrene, 3).

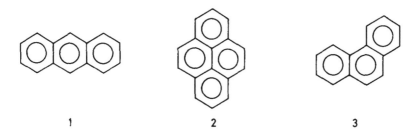

| 1 | 2 | 3 |

The theoretically possible number of PAH is tremendous. Table 1 shows the number of possible isomers which consist exclusively of six-membered rings, as a function of the number (n) of rings.[1] There are 683,101 theoretical possibilities for the linking of 12 rings alone. PAH may also contain five-membered rings, as in fluorene and fluoranthene, giving rise to a further great number of different PAH and structural isomers. In addition, alkyl groups may be attached to various positions of the unsubstituted parent PAH. These principal structural elements are encountered in mixed systems in an almost unlimited number of permutations. This large number of different PAH and isomers calls for a systematic, unambiguous nomenclature.

Nowadays, the proposals given by the International Union of Pure and Applied Chemistry (IUPAC) are generally used. Details of the rules for nomenclature of PAH have been published in *The Ring Index*.[2] A short summary of the IUPAC rules is given below.

1. A limited number of base components are given trivial names, e.g., anthracene, pyrene.
2. The peripheral numbering is dictated by placing the PAH so that the maximum number of rings lie in a horizontal row, and as many rings as possible are above and to the right of the horizontal row (see Figure 1). The system is then numbered in a clockwise direction starting with the carbon atom not engaged in ring fusion in the most counterclockwise position of the uppermost ring which is farthest to the right. Atoms common to two rings are not numbered.
3. The names of structures which have no trivial name are obtained by prefixing the name of a base component (which should contain as many rings as possible) with a designation of the other component, e.g., benzo(a)pyrene, benzo(b)fluoranthene (Figure 1).

Isomers are then distinguished by lettering the peripheral sides of the base component, beginning with ''a'' for side 1-2, ''b'' for side 2-3, etc. as illustrated in Figure 1. The letters are placed within parentheses between the two components, e.g., benzo(a)fluorene, dibenz(a,h)anthracene. If necessary, the numbering of carbon atoms of the smaller component

Table 1
NUMBER OF PAH WITH SIX-MEMBERED RINGS

n	Kata-annellated PAH	Peri-condensed PAH	Σ
1	1	0	1
2	1	0	1
3	2	1	3
4	5	2	7
5	12	10	22
6	37	45	82
7	123	210	333
8	446	1002	1448

From Zander, M., *Handbook of Polycyclic Aromatic Hydrocarbons*, Bjørseth, A., Ed., Marcel Dekker, New York, 1983, chap. 1. With permission.

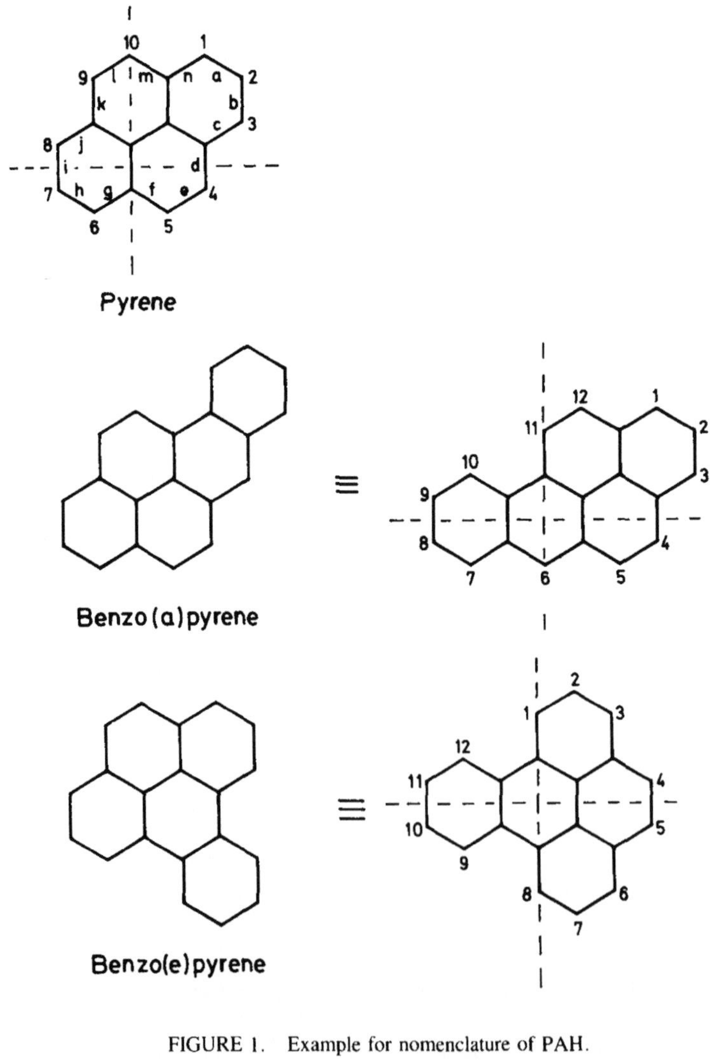

Pyrene

Benzo(a)pyrene

Benzo(e)pyrene

FIGURE 1.   Example for nomenclature of PAH.

participating in the fusion are placed within the parentheses before the side designation of the larger component, e.g., indeno(1, 2, 3-cd)pyrene.

As discussed earlier, high-temperature pyrolysis or incomplete combustion of organic materials are the main sources of PAH emission. These processes lead only to a limited number of specific, thermodynamically stable PAH which may be released into the environment. The most frequently occurring PAH from such pyrolytic processes are the unsubstituted parent PAH. Table 2 lists the name, structure, molecular weight, melting point, and boiling point of parent PAH (with 3 or more rings) most frequently found in environmental samples.

II. PHYSICAL PROPERTIES

The physical and spectroscopic properties of PAH are dominated by the size and topology of the π-electron system.[1] Here "size" is understood to mean the number of carbon centers, i.e., number of π-electrons, while "topology" denotes the type of ring linkage. Correlations among size, topology, and properties of PAH have been established in a number of theoretical models.[1]

All entirely unsaturated PAH are solids at ambient temperatures, and their boiling points are markedly higher than those of the n-alkanes of the same carbon number. The vapor pressures of the PAH at 25°C in Table 2 vary from 1.6×10^{-2} Pa for phenanthrene to 2×10^{-10} Pa for coronene.[3,4] This property has considerable impact on the amount of PAH that is adsorbed on particulate matter in the atmosphere and retained on particulate matter during collection of air samples on filters.

Table 3 shows the particulate/vapor distribution for the main PAH components in an ambient air sample.[5]

The collection efficiency of three-ring PAH from ambient air using glass fiber filters is usually lower than 5%.[5,6] At ambient summer temperatures, and during prolonged sampling even five-ring compounds such as benzo(a)pyrene may show considerable losses from the sampling filter.[3]

The extended π-electron system of PAH leads to strong absorption of visible or ultraviolet (UV) radiation which gives characteristic absorption and fluorescence spectra. The processes occurring when UV or visible light is absorbed by PAH are illustrated in Figure 2. The absorption of a light quantum promotes the electrons in the lowest vibrational level of the singlet ground state, S_0, to various vibrational levels of higher electronic states, S_2, S_2, etc. This process takes about 10^{-15} sec. The probability of absorption of discrete quanta is governed by the selection rules of quantum mechanics. When a π-bonding electron of a PAH absorbs energy, it is promoted to a π^*-antibonding orbital, often with a high molar absorptivity. The energy absorbed by the molecule can be dissipated by a number of photophysical or photochemical processes. The molecule relaxes rapidly by radiationless transfer of energy to the lowest vibrational levels of the first excited state, S_1. This is called internal conversion. From S_1 the molecule may either:

1. Return to the ground state S_0 by radiationless internal conversion
2. Return to S_0 by emission of light-producing fluorescence
3. Be converted to excited vibrational levels of the lowest triplet state T_1. After internal conversions to the lowest vibrational level of T_1, the molecule may undergo radiationless transition to the ground state S_0, or may emit the remaining energy as light-producing phosphorescence. Since T_1 has a lower energy than S_1. The wavelength of phosphorescence for a given molecule is longer than that of fluorescence.

The lifetime of S_1 is short, approximately 10 to 100 nsec for PAH, and process 1 is

Table 2
A LIST OF PAH — THEIR STRUCTURE, MOLECULAR WEIGHT,
MELTING POINT, AND BOILING POINT

Structure	IUPAC nomanclature (synonyms)	Molecular weight	Melting point (°C)	Boiling point (°C)[760]
	Dibenz[a,i] anthracene 1,2:6,7-Dibenzanthracene 1,2-Benzonaphthacene Isopentaphene	278.36	264	
	Dibenz[a,j] anthracene 1,2:7,8-Dibenzanthracene α,α'-Dibenzanthracene Dinaphthanthracene	278.36	197	
	Benzo[b] chrysene 1,2:6,7-Dibenzophenanthrene 3,4-Benzotetraphene Naphtho-2',1':1,2-anthracene	278.36	294	
	Picene Dibenzo[α,i] phenanthrene 3,4-Benzochrysene 1,2:7,8-Dibenzophenanthrene	278.36	368	519
	Benzo[ghi] perylene 1,12-Benzoperylene	276.34	278	
	Anthanthrene Dibenzo[def,mno] chrysene	276.34	264	
	Coronene Hexabenzobenzene	300.36	439	525
	Dibenzo[a,e] pyrene Dibenzo[a,e] pyrene	302.38	233	
	Benz[a] anthracene 1,2-Benzanthracene Tetraphene 2,3-Benzophenanthrene Naphthanthracene	228.30	167	435 sub
	Triphenylene 9,10-Benzophenanthrene Isochrysene	228.30	199	425

Key: d = decomposes ; sub = sublimes

Table 2 (continued)
A LIST OF PAH — THEIR STRUCTURE, MOLECULAR WEIGHT,
MELTING POINT, AND BOILING POINT

Structure	IUPAC nomenclature (synonyms)	Molecular weight	Melting point (°C)	Boiling point (°C) [760]
	Chrysene 1,2-Benzophenanthrene Benzo[a]phenanthrene	228.30	256	448
	Benzo[b]fluoranthene 2,3-Benzofluoranthene 3,4-Benzofluoranthene Benz[e]acephenanthrylene	252.32	168	481
	Benzo[j]fluoranthene 7,8-Benzofluoranthene 10,11-Benzofluoranthene	252.32	165	480
	Benzo[k]fluoranthene 8,9-Benzofluoranthene 11,12-Benzofluoranthene	252.32	216	480
	Benzo[e]pyrene 4,5-Benzpyrene 1,2-Benzopyrene	252.32	179	493
	Benzo[a]pyrene 1,2-Benzpyrene 3,4-Benzopyrene Benzo[def]chrysene	252.32	178	496
	Perylene peri-Dinaphthalene	252.32	274	
	Indeno[1,2,3-cd]pyrene o-Phenylenepyrene	276.34	164	
	Dibenz[a,c]anthracene 1,2:3,4-Dibenzanthracene Naphtho-2',3',:9,10-phenanthrene	278.36	206	
	Dibenz[a,h]anthracene 1,2:5,6-Dibenzanthracene	278.36	267	
	Acenaphthylene	152.21	93	270 d

Table 2
A LIST OF PAH — THEIR STRUCTURE, MOLECULAR WEIGHT, MELTING POINT, AND BOILING POINT

Structure	IUPAC nomenclature (synonyms)	Molecular weight	Melting point (°C)	Boiling point (°C) [760]
	Acenaphthene Naphthyleneethylene	154.21	96	279
	Fluorene 2,3-Benzindene Diphenylenemethane	166.23	117	295
	Phenanthrene o-Diphenyleneethylene	178.24	100	340
	Anthracene	178.24	218	342
	Fluoranthene Idryl 1,2-Benzacenaphthene Benzo[jk]fluorene Benz[a]acenaphthylene	202.26	111	375
	Pyrene Benzo[def]phenanthrene	202.26	156	404
	Benzo[a]fluorene 11H-Benzo[a]fluorene 1,2-Benzofluorene Chrysofluorene	216.29	190	413
	Benzo[b]fluorene 11H-Benzo[b]fluorene 2,3-Benzofluorene Isonaphthofluorene	216.29	209	402
	Benzo[ghi]fluoranthene	226.28	149	432
	Benzo[c]phenanthrene 3,4-Benzophenanthrene	228.30	66	

Table 3
DISTRIBUTION (%) OF SELECTED
PAH ON FILTER AND
POLYURETHANE FOAM IN AMBIENT
AIR SAMPLE COLLECTED IN OSLO,
NORWAY

Compound	Particles	Gas phase
Naphthalene	0	100
Biphenyl	0	100
Fluorene	0	100
Phenanthrene	2	98
Anthracene	3	97
2-Methylanthracene	29	71
Fluoranthene	52	48
Pyrene	56	44
Benzo(a)fluorene	71	29
Benz(a)anthracene	95	5
Chrysene/triphenylene	97	3
Benzo(a)pyrene	100	0
Benzo(a)pyrene	100	0
Perylene	100	0
Indeno(1,2,3-cd)pyrene	100	0
Benzo(ghi)perylene	100	0
Coronen	100	0

Adapted from Thrane, K. and Mikalsen, A., *Atmos. Environ.*, 15, 909, 1981.

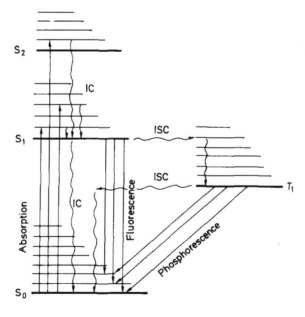

FIGURE 2. Simplified energy-level diagram for PAH, IC, internal conversion: ISC, intersystem crossing.

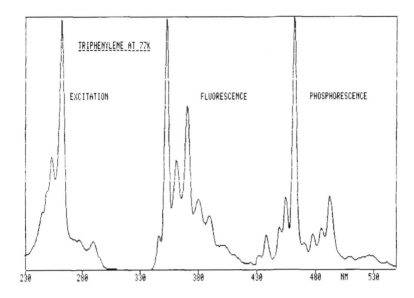

FIGURE 3.   Relative position of excitation, fluorescence, and phosphorescence bands of triphenylene. (From Rhys Williams, A. T., *Fluorescence Detection in Liquid Chromatography,* Perkin-Elmer, Beaconsfield, 1980. With permission.)

generally not important. For anthracene dissolved in ethanol at 20°C, about 30% of the excited singlet disappears by the $S_1 \rightarrow S_0$ fluorescence process and 70% by the $S_1 \rightarrow T_1$ intersystem crossing process. As the emission of fluorescence takes place from the lowest vibrational level of S_1, the emitted light is usually of lower energy than the absorbed, and the fluorescence bands will occur at longer wavelength than the absorption bands.[7] This is demonstrated in Figure 3, showing the relative position of absorption, fluorescence, and phosphorescence bands of triphenylene. The lifetime of the triplet state T_1 can be considerably longer than that observed from 10^{-4} to 100 sec. Since molecules occupy the triplet state for relatively long periods, they can easily lose their energy by collision. Therefore, phosphorescence at room temperature is a rare phenomenon. However, by cooling the sample to low temperatures, phosphorescence is observed in many compounds.

As most of the PAH absorb light at wavelengths found in sunlight (> 300 mm), it is likely that they may photochemically react by direct excitation. Indeed, photo-oxidation is probably one of the most important processes in the removal of PAH from the atmosphere.[8,9]

III. ATMOSPHERIC FORMS OF PAH

Because of the high melting point and low vapor pressure of PAH with four and more rings, these compounds are generally linked with aerosols. It is believed that airborne PAH exist either as particles of relatively pure material, or are adsorbed to particulate matter such as soot, dust, or other material.[8] However, the exact physical state of the various PAH in the atmosphere is not known.

Some information is available on the relationship of PAH to particle size, and this has particular significance with respect to the entry of PAH into the respiratory tract, and the deposition and retention in the organism. The aerodynamic size of the particles will determine how much and where they are deposited in the respiratory tract. Relatively large particles (diameter > 10 μm) are deposited in the nasal and oral cavities. Smaller particles will pass into the trachea-bronchial regions, and particles smaller than 2 μm may reach the alveolar compartment and be deposited there.[10,11] The actual deposition efficiencies within the respiratory tract, however, will vary considerably.

Table 4
PAH (μg) ON PARTICLES AS A FUNCTION OF PARTICLE SIZE AT A COKE OVEN BATTERY TOP

	Particle size				
	>15 μm	7—15 μm	3—7 μm	0.9—3 μm	<0.9 μm
Phenanthrene	0.5	0.5	0.8	2.0	*
Anthracene	0.3	0.2	0.6	0.7	*
Methylphenanthrene/methylanthracene[a]	—	—	0.4	1.3	—
Fluoranthene	—	—	4.2	10.7	2.0
Dihydrobenzo(a,b)fluorene	—	—	0.6	2.1	—
Pyrene	—	—	4.3	12.3	1.3
Benzo(a)fluorene	—	—	3.5	9.1	0.5
Benzo(b)fluorene	—	—	2.6	7.2	0.3
Benzo(c)phenanthrene	—	—	2.0	5.2	0.8
Benz(a)anthracene	0.03	0.08	3.3	9.0	4.0
Chrysene/triphenylene	0.06	0.10	5.8	13.2	1.8
Benzo(b + j + k)fluoranthene	0.05	0.40	4.8	10.6	7.2
Benzo(e)pyrene	0.03	0.14	3.4	8.9	0.5
Benzo(a)pyrene	0.02	0.12	5.2	12.1	1.4
Perylene	—	—	1.3	2.9	—
o-Phenylenepyrene	—	—	3.8	7.2	0.5
Benzo(ghi)perylene	—	—	5.2	7.8	1.3
Anthranthrene	—	—	1.2	3.2	0.3
Coronene	—	—	5.9	6.3	0.5
Dibenzopyrene	—	—	0.9	1.5	0.3

Note: *, detectable; —, not detected.

[a] The isomer not determined.

From Bjørseth, A., Bjørseth, O., and Fjeldstad, P. E., *Scand. J. Work Environ. Health,* 4, 224, 1978. With permission.

A number of studies have been reported relating to the distribution of atmospheric PAH with respect to particle size.[12-17] The PAH in urban aerosols predominate in small particles with an aerodynamic diameter smaller than 0.5 μm. The situation seems to be different in workplace environments.[12,18,19] Table 4 shows the distribution of PAH on particles sampled at the top of a coke oven battery as a function of particle size.[19] Only 10% of the PAH is found on particles smaller than 0.9 μm, while more than 60% is found on particles between 0.9 and 3 μm. This is in accordance with results from Miguel and Rübenich[12] who found 68% of BaP on particles from a coke oven sample with an aerodynamic diameter between 0.5 and 4.0 μm. Although larger particles seem to be observed in the atmospheres of workplaces, such as coke ovens, compared to ambient air, still a negligable amount of the total PAH is connected to particles classified as not respirable, i.e., particles with diameters larger than 10 μm.

IV. CHEMICAL REACTIVITY

Reactions of PAH appear to play a major role in the removal of PAH from the atmosphere.[8] Recent evidence suggests that in some instances the transformation products formed may even exhibit a much greater biological activity than the parent PAH.[20-23] Thus, transformation reactions may be important in the consideration of airborne PAH as a health hazard. The atmospheric stability of different PAH varies greatly and is dependent upon such factors as molecular structure of the PAH, the amount of available light, the presence of co-pollutants such as O_3, NO, and SO_2, and on the characteristics of the particles to which the PAH are adsorbed.

A. Oxidation Reactions

Two general types of oxidation reactions appear to be possible:[8] (1) photochemically induced oxidation, and (2) oxidation by chemical oxidants such as ozone or free radicals.

PAH have strong absorption of UV radiation at wavelengths longer than 300 μm, and most PAH are readily photo-oxidized by air and sunlight. The earliest study of PAH photochemical degradation was conducted by Falk and co-workers.[24] A striking result was the higher reactivity, in light and air, for PAH in the pure form vs. the same PAH adsorbed on soot. This was explained by the hypothesis that adsorption on a highly porous particle may provide some protection from photooxidation.

There have been several other reports of destruction of PAH adsorbed on soot or smoke.[25-27] Most of the destructions require light, although they are accelerated by synthetic photochemical smog.[24,25] It is difficult to calculate specific rates from the data presented, but exposure for roughly 40 min to light of one fourth the intensity of noon sunlight caused a 35 to 65% loss of BaP in airborne smoke samples in an irradiated flow reactor.[26,27] These studies imply that the chemical half-life of BaP in the presence of light and photochemical oxidants will be short, probably in the range of hours to days.

Relatively few studies have attempted to identify the photooxidative products. Masuda and Kuratsune[28] and Inscoe[29] have identified quinones of BaP and pyrene. Fox and Olive[30] have shown that anthracene, when dispersed into atmospheric particulate matter, is photooxidized by air and sunlight to a variety of products, including the 9, 10-endoperoxide and various quinones. Recently, a great number of polycyclic aromatic ketones have been identified in combustion emissions as well as in samples from an aluminum plant work atmosphere and urban air.[31]

The nonphotochemical oxidation of PAH involve, for example, reactions with ozone and free radicals. Ozone reacts readily with PAH. Studies by Lane and Katz[9] have shown high reactivity of BaP on petri dishes when it is exposed to sub-ppm levels of ozone in air. Irradiation did not seem to significantly affect the reaction rate. They also observed that the reactivity of the PAH varies strongly with their structure. Experiments by Pitts and co-workers[32] have confirmed and extended this work. For the atmospheric reaction of BaP with 0.2 ppm O_3 they found conversion yields of 50% after only 1 hr and 80% after 4 hr.

Reactions of free radicals, such as the hydroxyl radical, with aromatic hydrocarbons have been investigated only in a few cases.[33-35] From the study of simple model systems one might expect the formation of hydroxy derivatives of PAH and ring-opening oxidation products.[33] The former could react further, for example, to quinones.

Korfmacher et al.[36] have found that certain PAH are rapidly oxidized in the dark to the corresponding ketones and quinones when adsorbed onto coal fly ash or charcoal. As it has been observed that soot particles contain free radicals,[37,38] the spontaneous oxidation of certain adsorbed PAH might proceed by a radical pathway. Recently, Alfheim et al.[39] studied the thermal transformation of PAH adsorbed on secondary alumina from a dry scrubbing system of an aluminum plant. The amount of extractable PAH decreased rapidly during the thermal treatment of the alumina. No significant amounts of PAH were driven off during heating of the secondary alumina for 30 min to 340°C and PAH was no longer detectable in the extracts after heating. It was suggested that the PAH had been oxidized to polar compounds which were firmly bound to the surface of the alumina.

B. Reactions with Oxides of Nitrogen and Sulfur

During the last few years several papers have been published concerning reactions between PAH and gases such as NO, NO_2, SO_3, and SO_3 which may be formed simultaneously with

Table 5
REACTIVITY SCALE OF PAH IN ELECTROPHILIC REACTIONS

Reactivity class	PAH compounds
I	Benzo(a)tetracene, dibenzo(a,h)pyrene, tetracene, pentacene
II	Anthracene, anthanthrene, benzo(a)pyrene, perylene, dibenzo(a,l)pyrene, dibenzo(a,i)pyrene, dibenzo(a,c)tetracene
III	Pyrene, benz(a)anthracene, benzo(g,h,i,)perylene, cyclopenta(c,d)pyrene, benzo(g)chrysene, picene, dibenzo(a,e)pyrene
IV	Benzo(e)pyrene, chrysene, coronene, dibenzanthracenes, benzo(c)phenanthrene, benzo(c)chrysene, dibenzo(e,j)pyrene
V	Phenanthrene, fluoranthene, triphenylene, benzofluoranthenes, indeno(1,2,3-cd)pyrene

From Nielsen, T., Nordic PAH-project, Report No. 10, Central Institute for Industrial Research, Oslo, 1981.

PAH during combustion processes. Pitts et al.[40] have demonstrated that BaP and perylene deposited on glass fiber filters will form nitro derivatives of PAH by exposure to NO_2 and traces of nitric acid. Studies by Ramdahl et al.[41] indicate that the reaction of NO_2 with particle-adsorbed PAH in air is strongly dependent on the type of carrier. The transformation of PAH decreased in this order: silica, alumina, and charcoal. Thus, PAH seems to be stabilized and protected by carbonaceous particles, and nitration of PAH in the atmosphere may be slow.

Several of the nitro-PAH exhibit a high, direct-acting mutagenic activity in *Salmonella* tests and some are shown to be carcinogenic in animals.[40,42,43] Nitro-PAH have also been identified in several environmental samples such as ambient air particulates,[44] diesel exhaust,[45] and in the work atmosphere of an aluminum smelter.[46]

As a result of the sulfur content of fossil fuels, sulfur oxides are formed by combustion. The original species formed are largely SO_2, which, however, may be easily oxidized to SO_3, and sulfate, especially when adsorbed on particles.[47] It has been demonstrated that PAH react with SO_3[48] and with SO_2 under the influence of light[25] or when adsorbed to soot particles.[49] Some PAH sulfonic acids, like pyrene- and BaP-sulfonic acid, have been identified as reaction products in laboratory experiments.[49] However, thus far such sulfonated PAH have not been found in real samples from ambient air or workplace atmospheres. These acidic compounds are only slightly soluble in usual extraction solvents, and may therefore have been overlooked in previous studies of airborne PAH.

It is reasonable to assume that the sulfur and nitrogen oxides form acids with water vapor in ambient air, or with water adsorbed on the particles. Thus, the transformation reactions are likely to be of electrophilic character. Based on this assumption, Nielsen[50] determined the decomposition rates of PAH in electrophilic aromatic reactions and proposed a classification of reactivity of PAH. The reactivity scale is given in Table 5. The PAH have been divided into five groups. The most reactive ones are in the group with the lowest number, and the most stable ones are in the group with the highest number.

V. SUMMARY AND CONCLUSION

PAH with four and more rings are generally associated with particulate matter when detected in ambient air and workplace atmospheres. The major portion of PAH in ambient air is found with particles smaller than 0.5 μm, while PAH from workplace atmospheres seem to be associated with somewhat larger particles, though still in the respirable range (<10 μm).

Most properties of PAH, such as UV absorption, fluorescence, and chemical reactivity, depend on the size and topology of the π-electron system. Experimental chemical and adsorbed. There is evidence that they are degraded in the atmosphere by photooxidation, by reaction with atmospheric oxidants and sulfur and nitrogen oxides. These reactions are important with respect to the fate of PAH. However, it has been demonstrated that transformation products of PAH may also exhibit a higher biological activity than the original PAH.

The instability of PAH resulting from volatility, photosensitivity, and chemical transformation has an important bearing on the sampling of particulate PAH. Considerable losses of PAH during filter sampling might occur, arising from saturated vapor concentrations or degradation reactions.

REFERENCES

1. **Zander, M.**, Physical and chemical properties of polycyclic aromatic hydrocarbons, in *Handbook of Polycyclic Aromatic Hydrocarbons*, Bjørseth, A., Ed., Marcel Dekker, New York, 1983, chap. 1.
2. **Patterson, A. M., Capell, L. T., and Walker, D. F.**, *The Ring Index. A List of Ring Systems Used in Organic Chemistry*, 2nd ed., American Chemical Society, Washington, D. C., 1960, 1425.
3. **Pupp, C., Lao, R. C., Murray, J. J., and Pottie, R. F.**, Equilibrium vapour concentrations of some polycyclic aromatic hydrocarbons, As_4O_6 and SeO_2 and the collection efficiencies of these air pollutants, *Atmos. Environ.*, 8, 915, 1974.
4. **Sonnefeld, W. J., Zoller, W. H., and May W. E.**, Dynamic coupled-column liquid chromatographic determination of ambient temperature vapor pressure of polynuclear aromatic hydrocarbons, *Anal. Chem.*, 55, 275, 1983.
5. **Thrane, K. and Mikalsen, A.**, High-volume sampling of airborne polycyclic aromatic hydrocarbons using glass fibre filters and polyurethane foam, *Atmos. Environ.*, 15, 909, 1981.
6. **Cautreels, W. and van Cauwenberghe, K.**, Experiments on the distribution of organic pollutants between airborne particulate matter and the corresponding gas phase, *Atmos. Environ.*, 12, 1133, 1978.
7. **Rhys Williams, A. T.**, *Fluorescence Detection in Liquid Chromatography*, Perkin-Elmer, Beaconsfield, U.K., 1980.
8. Committee on Biological Effects of Atmospheric Pollutants, *Particulate Polycyclic Organic Matter*, National Academy of Sciences, Washington, D.C., 1972.
9. **Lane, D. A. and Katz, M.**, The photomodification of benzo(a)pyrene, benzo(b)fluoranthene and benzo(k)fluoranthene under simulated atmospheric conditions, in *Advances in Environmental Science and Technology*, Vol. 9, *Fate of Pollutants in the Air and Water Environments*, Suffets, I. H., Ed., Interscience, New York, 1977, 137.
10. **Pott, F. and Oberdörster, G.**, Intake and distribution of PAH, in *Environmental Carcinogens: Polycyclic Aromatic Hydrocarbons*, Grimmer, G., Ed., CRC Press, Boca Raton, Fla., 1983, 130.
11. International Radiological Protection Commission, Task Group on Lung Dynamics, Deposition of retention models for internal dosimetry of the human respiratory tract, *Health Phys.*, 12, 173, 1966.
12. **Miguel, A. H. and Rübenich, L. M. S.**, Submicron size distributions of particulate polycyclic aromatic hydrocarbons in combustion source emissions, in *Polynuclear Aromatic Hydrocarbons: Chemistry and Biological Effects*, Bjørseth, A. and Dennis, A. J., Eds., Battelle Press, Columbus, Ohio, 1980, 1077.
13. **Albagli, A., Oja, J., and Dubois, L.**, Size distribution pattern of polycyclic aromatic hydrocarbons in airborne particulates, *Environ. Lett.*, 6, 241, 1974.
14. **Kertész-Saringer, M., Mészaros, E., and Varkonyi, T.**, On the size distribution of benzo(a)pyrene-containing particles in urban air, *Atmos. Environ.*, 5, 429, 1971.
15. **Miguel, A. H. and Friedlander, S. K.**, Distribution of benzo(a)pyrene and coronene with respect to particle size in Pasadena aerosols in the submicron range, *Atmos. Environ.*, 12, 2407, 1978.
16. **Pierce, R. C. and Katz, M.**, Dependency of polynuclear aromatic hydrocarbon content on size distribution of atmospheric aerosols, *Environ. Sci. Technol.*, 9, 347, 1975.
17. **Starkey, R. and Warpinski, J.**, Size distribution of particulate benzo(a)pyrene, *J. Environ. Health*, 36, 503, 1974.
18. **Broddin, G., Van Vaeck, L., and Van Cauwenberghe, K.**, On the size distribution of polycyclic aromatic hydrocarbon containing particles from a coke oven emission source, *Atmos. Environ.*, 11, 1061, 1977.

19. **Bjørseth, A., Bjørseth, O., and Fjeldstad, P. E.,** Polycyclic aromatic hydrocarbons in the work atmosphere. II. Determination in a coke plant, *Scand. J. Work Environ. Health,* 4, 224, 1978.

20. **Löfroth, G.,** Comparison of the mutagenic activity from diesel and gasoline powered motor vehicles to carbon particulate matter, in *Application of Short-Term Bioassays in the Analysis of Complex Environmental Mixtures,* Vol. 2, Plenum Press, New York, 1980, 319.

21. **Teranishi, K., Hamada, K., and Watanabe, H.,** Mutagenicity in *Salmonella typhimurium* mutants of the benzene-soluble organic matter derived from airborne particulate matter and its five fractions, *Mutation Res.,* 56, 273, 1978.

22. **Alfheim, I., Becher, G., Hongslo, J. K., and Ramdahl, T.,** Mutagenicity testing of high performance liquid chromatography fractions from wood stove emission samples using a modified *Salmonella* assay requiring smaller sample volumes, *Environ. Mutagenesis,* 6, 91, 1984.

23. **Pitts, J. N., Jr., Grosjean, D., Mischke, T. M., Simmon, V. F., and Poole, D.,** Mutagenic activity of airborne particulate organic pollutants, *Toxicol. Lett.,* 1, 65, 1977.

24. **Falk, H. L., Markul, I., and Kotin, P.,** Aromatic hydrocarbons. IV. Their fate following emission into the atmosphere and experimental exposure to washed air and synthetic smog, *A.M.A. Arch. Ind. Health,* 13, 13, 1956.

25. **Falk, H. L., Kotin, P., and Thompson, S.,** Inhibition of carcinogenesis. The effect of polycyclic aromatic hydrocarbons and related compounds, *Arch. Environ. Health,* 9, 169, 1964.

26. **Tebbens, B. D., Thomas, J. F., and Mukai, M.,** Fate of arenes incorporated with airborne soot, *Am. Ind. Hyg. Assoc. J.,* 27, 415, 1966.

27. **Thomas, J. F., Mukai, M., and Tebbens, B. D.,** Fate of airborne benzo(a)pyrene, *Environ. Sci. Technol.,* 2, 33, 1968.

28. **Masuda, Y. and Kuratsune, M.,** Photochemical oxidation of benzo(a)pyrene, *Air Water Pollut. Int. J.,* 10, 805, 1966.

29. **Inscoe, M. N.,** Photochemical changes in thin-layer chromatograms of polycyclic aromatic hydrocarbons, *Anal. Chem.,* 36, 2505, 1964.

30. **Fox, M. A. and Olive, S.,** Photoxidation of anthracene on atmospheric particulate matter, *Science,* 205, 582, 1979.

31. **Ramdahl, T.,** Polycyclic aromatic ketones in environmental samples, *Environ. Sci. Technol.,* 17, 666, 1983.

32. **Pitts, J. N., Lokensgard, D. M., Ripley, P. S., Van Cauwenberghe, K. A., Van Vaeck, L., Shaffer, S. D., Thill, A. J., and Belser, W. L., Jr.,** "Atmospheric" epoxidation of benzo(a)pyrene by ozone: formation of the metabolic benzo(a)pyrene-4,5-oxide, *Science,* 210, 1347, 1980.

33. **Atkinson, R., Carter, W. P. L., Darnall, K. R., Winer, A. M., and Pitts, J. N., Jr.,** A smog chamber and modelling study of the gas phase NO-air photo-oxidation of toluene and cresols, *Int. J. Chem. Kinet.,* 12, 779, 1980.

34. **Carter, W. P. L., Winer, A. M., and Pitts, J. N., Jr.,** Major atmospheric sink for phenol and cresols. Reaction with nitrate radical, *Environ. Sci. Technol.,* 15, 829, 1981.

35. **Kenley, R. A., Davenport, J. E., and Hendry, D. G.,** Hydroxyl radical reactions in the gasphase. Products and pathways for the reaction of OH with toluene, *J. Phys. Chem.,* 82, 1095, 1978.

36. **Korfmacher, W. A., Natusch, D. F. S., Taylor, D. R., Mamantov, G., and Wehry, E. L.,** Oxidative transformations of polycyclic aromatic hydrocarbons adsorbed on coal fly ash, *Science,* 207, 763, 1980.

37. **Bennett, J. E., Ingram, D. J. E., and Tapley, J. G.,** Paramagnetic resonance from broken carbon bonds, *J. Chem. Phys.,* 23, 215, 1955.

38. **Winslow, F. H., Baker, W. O., and Yager, W. A.,** Odd electrons in polymer molecules, *J. Am. Chem. Soc.,* 77, 4751, 1955.

39. **Alfheim, I., Kveseth, K., Ramdahl, T., and Rob, J.,** Thermal oxidation of polycyclic aromatic hydrocarbons adsorbed on alumina, *Scand. J. Work Environ. Health,* 11, 439, 1985.

40. **Pitts, J. N., Jr., Van Cauwenberghe, K. A., Grosjean, D., Schmid, J. P., Fitz, D. R., Belser, W. L., Jr., Knudson, G. B., and Hynds, P. M.,** Atmospheric reactions of polycyclic aromatic hydrocarbons: facile formation of mutagenic nitro derivatives, *Science,* 202, 515, 1978.

41. **Ramdahl, T., Bjørseth, A., Lokensgard, D. M., and Pitts, J. N., Jr.,** Nitration of polycyclic aromatic hydrocarbons adsorbed to different carriers in a fluidized bed reactor, *Chemosphere,* 13, 527, 1984.

42. **Löfroth, G., Hefner, E., Alfheim, I., and Møller, M.,** Mutagenicity in photocopies, *Science,* 209, 1037, 1980.

43. **Rosenkranz, H. S., McCoy, E. C., Sanders, D. R., Butler, M., Kiriazides, D. K., and Mermelstein, R.,** Nitropyrenes: isolation, identification, and reduction of mutagenic impurities in carbon black and toners, *Science,* 209, 1039, 1980.

44. **Ramdahl, T., Becher, G., and Bjørseth, A.,** Nitrated polycyclic aromatic hydrocarbons in urban air particles, *Environ. Sci. Technol.,* 16, 861, 1982.

45. **Schuetzle, D., Riley, T. L., Prater, T. J., Harvey, T. M., and Hunt, D. F.,** Analysis of nitrated polycyclic aromatic hydrocarbons in diesel particulates, *Anal. Chem.,* 54, 265, 1982.

46. **Oehme, M., Manø, S., and Stray, H.,** Determination of nitrated polycyclic hydrocarbons in aerosols using capillary gas chromatography combined with different electron capture detection methods, *J. High Resol. Chromatogr. Chromatogr. Commun.,* 5, 417, 1982.

47. **Charlston, R. J., Covert, D. S., Larson, T. V., and Waggoner, A. P.,** Chemical properties of tropospheric sulfur aerosols, *Atmos. Environ.,* 12, 39, 1978.

48. **Hughes, M. M., Natusch, D. F. S., Taylor, D. R., and Zeller, M. V.,** Chemical transformation of particulate polycyclic organic matter, in *Polynuclear Aromatic Hydrocarbons: Chemistry and Biological Effects,* Bjørseth, A. and Dennis, A. J., Eds., Battelle Press, Columbus, Ohio, 1980, 1.

49. **Jäger, J. and Rakovic, M.,** Sulfur dioxide-induced qualitative changes in polycyclic aromatic hydrocarbons adsorbed on solid carriers, *J. Hyg. Epidemiol. Microbiol. Immunol.,* 18, 137, 1974.

50. **Nielsen, T.,** A study of the reactivity of polycyclic aromatic hydrocarbons, Nordic PAH-project, Rep: No. 10, Central Institute for Industrial Research, Oslo, 1981.

Chapter 4

SOURCES AND EXPOSURE

I. SOURCES OF PAH IN THE WORK ENVIRONMENT

PAH in the workplace atmosphere may originate from two major processes: (1) evaporation during heating of PAH-containing matter, and (2) formation by pyrolysis or incomplete combustion. Coal tar products, derived from the carbonization of bituminous coal, are the most important sources of PAH emissions in the work environment.[1,2] In many industrial processes these products are heated to high temperatures, thereby releasing PAH into the work atmosphere. Furthermore, at high temperatures, formation of PAH by pyrolysis may occur. In particular, coking operations and aluminum production have been recognized as major contributors to the airborne PAH burden and are documented occupational hazards.

Table 1 summarizes the approximate BaP content of some industry-related materials. Some of the most important industries with PAH problems are described below.

A. Coke Production[3,4]

1. Historical Perspectives

Until the 1700s, charcoal was used as both a fuel and a reducing agent to produce iron from its various ores. Since the mid-1700s the use of coke, which is the residue of destructive distillation of coal, had been generally adopted. Coke was produced in two types of ovens. The first oven constructed was the "beehive", a term which describes its shape. The coal was charged into the top of the oven and then leveled. The door on the side of the oven was then completely bricked. After the coal was coked, the brickwork was removed and the coke was sprayed with water to cool it, and then removed from the oven. In this type of oven, the volatiles were not collected and the emissions were allowed to escape into the atmosphere.

The second type of oven is the by-product, or slot oven, in which the volatilized products are recovered. By the 1850s these coke ovens, which could produce coke and recover the volatiles, had been developed. Today, the vast majority of coke is produced in this type of oven.

2. Coal Characteristics

Bituminous coal is used in the coking process. High-volatile bituminous coal is usually blended with low-volatile and/or medium-volatile coal. This blending of different coals provides a coke of sufficient quality and strength. The coals should also contain minimal amounts of both sulfur and ash.

3. Structure and Operation of a Coke Plant

A by-product coke battery usually consists of 10 to 100 ovens in parallel rows. A coke plant comprises from 1 to 12 batteries. A schematic view of a coke oven is given in Figure 1. The coke plant includes the coal-unloading station, a system of conveyors to transfer the coal to a storage bunker; a coke-quenching station, where the coal is cooled; a coke wharf, where the quenched coke is stored; a coke-screening station, where the coke is sized; and a car-loading station.

The individual by-product coke oven is basically composed of the coking chamber, heating flues, and a regenerative chamber. In the battery, the heating flues and coking chambers alternate so that a heat chamber is located on each side of a coking chamber. The regenerative chamber is located beneath the other two. The coking chamber is the place in which the

Table 1
APPROXIMATE BaP
CONTENT OF SOME
INDUSTRY-RELATED
MATERIALS

Materials	ppm BaP
Crude oils	
Petroleum	1
Shale-derived	3
Coal-derived	3
Petroleum products	
Gasoline	0.4
Motor oil, new	0.03
Motor oil, used	4
Diesel fuel	0.05
Asphalt	2
Petroleum pitch	2,000
Coal-derived products	
Creosote	200
Coal tar	3,000
Coal tar pitch	10,000

From Guerin, M. R., *Polycylic Hydrocarbons and Cancer*, Vol. 1, Gelboin, H. V. and Ts'o, P. O. P., Eds., Academic Press, New York, 1978, chap. 1. With permission.

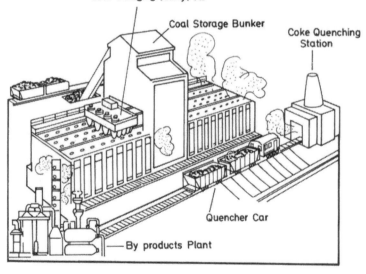

FIGURE 1.   Schematic view of a coke oven battery.

coal is destructively distilled into coke, the heating flues are where the fuel gas is burned, and the regenerative chambers contain the regenerations that permit separate control of preheated air for combustion to the individual, vertical heating-flue walls. There are doors on both ends of the refractory-bricked coking chamber. The ends are designated as the "push side", where the pushing ram enters to force the coked coal out of the oven, and the "coke side". A typical oven is 3 to 7 m high, 11 to 15 m long, and 42.5 to 50 cm broad.

The coking cycle begins when the coal is charged into the coking chamber through holes on top of the oven. A larry car, which has hoppers for the coal and travels on rails, brings the coal to the charging holes. After charging, the lids that cover the charging holes are replaced and luted to prevent escape of emissions into the atmosphere.

The temperature of the coal is raised over a period of 14 to 34 hr (depending on the type of coke to be made) to temperatures in excess of 1000°C. At this elevated temperature, destructive distillation of the coal occurs. The volatiles that are driven off are collected in the by-product recovery system, leaving a mass of hard coke in the chamber.

After coking, the doors on both sides of the oven are removed and a ram, located on the pusher machine, pushes coke out of the oven through a coke guide. The quench car collects the coke and conveys it to the quenching station, where the hot coke is cooled below its ignition point in air, usually by spraying water over it.

At the screening station, the coke is sized and sent to the loading station. When the doors of the oven are replaced, the coking cycle is ready to be repeated.

4. Major By-Products

There are three basic changes that occur in the coal as it is being coked, which correspond roughly to the following temperatures:

1. Less than 700°C: formation of breakdown products of coal, which are water, carbon monoxide, carbon dioxide, hydrogen sulfide, olefins, paraffins, hydroaromatics, phenolic compounds, and nitrogen-containing compounds
2. At 700°C: thermal reactions of the primary products resulting in formation of aromatic hydrocarbons and methane, evolution of hydrogen, and decomposition of the nitrogen-containing compounds to hydrogen cyanide, ammonia, and nitrogen
3. More than 700°C: production of hard coke by the removal of hydrogen

It has been estimated that 20 to 35% of the initial coal charged is evolved as vapors and gases.

Coke oven gas, which is used as a fuel, consists mainly of hydrogen, methane, ethane, carbon monoxide, carbon dioxide, gaseous hydrocarbons, hydrogen sulfide, ammonia, and nitrogen.

Coke oven gas tar, which condenses from the gas in the collector mains, consists of a complex mixture of liquid hydrocarbons, such as benzene, toluene, xylene, and solvent naphthas; nitrogen-containing compounds, such as pyridine; phenols; and creosote oil and coal tar pitch containing PAH.

Theoretically, the by-product process collects all volatile material given off during coking. However, this is not the case during the actual operation of a coke plant. The reasons that emissions escape include lack of engineering controls, structural defects in the battery, such as improper sealing of doors and charging lids, improper use of installed engineering controls, and improper working practices.

The emissions from a coke oven are complex mixtures of gases, liquids, and solid particles whose exact composition may vary considerably depending upon the type of coal used and the method of operation in the coking facility.[5] The particulate matter is a mixture of irregularly shaped coal and coke particles, and particles of tarry material. The airborne particulate matter on top of a coke battery has been shown to contain approximately 15 wt% PAH.[6]

5. Personnel

The average coke battery will have approximately 20 to 30 operating, maintenance, and supervisory personnel assigned to a crew per work shift. A coke battery is in operation 24 hr a day. The number of ovens pushed per shift is dependent on the coking cycle and the size of the battery. On average, approximately 30 ovens are pushed per shift.

The work crew of the battery is divided by area. The areas are topside and benchside, the latter being further divided into push and coke sides. Job classifications on the topside include larry car operator, lidman, luterman, relief, and tar chaser. On the push side of the bench are the pusher machine operator, benchmen, oven patchers, luterman (bench-side) and relief. Coke side job classifications are door-machine operator, quench car operator, door cleaners, and relief.

B. Aluminum Production[7,8]

1. Introduction

The basis for modern commercial production of aluminum was established in 1886 independently by both Hall in the U.S. and Héroult in France who developed a process for the electrolytic reduction of alumina (Al_2O_3) to produce aluminum. As alumina has a melting point close to 2000°C, the problem was how to render alumina into a molten state so that an electrolytic method could be used. This was accomplished by dissolving alumina in molten cryolite (Na_3AlF_5) which melts at a temperature below 1000°C. Cryolite can dissolve approximately 8% alumina. In the electrolysis, aluminum migrates to the cathode, and oxygen is released at the anode. This same process is used today and is referred to as the Hall-Héroult process.

2. Electrolytic Reduction

The electrolytic reduction is carried out in carbon-lined steel shells which serve as cathodes. Carbon anodes are immersed in the molten mixture of alumina and cryolite whose temperature is between 950 and 970°C. The cells are commonly known as pots. During electrolysis, aluminum is formed at the cathode and oxygen at the anode. The oxygen gradually consumes the carbon anode by forming carbon monoxide and carbon dioxide. The process is summarized in the following equation:

$$2\ Al_2O_3\ +\ 3\ C\ \xrightarrow{\text{electrolysis}}\ 4\ Al\ +\ 3\ CO_2$$

Usually, pots are connected in series in a potroom. A plant might have 250 pots, producing 100,000 tonnes of aluminum per year. There are two basic types of pots in use — prebake and Söderberg — but within each category there may be some differences in pot design. A plant may operate more than one type of pot.

a. Prebake Anode Process

Typical prebake pots are shown in Figures 2a and 2b. For prebake pots, the anodes are fabricated in a carbon plant which is separated from the potrooms. In the green-carbon section, coke (usually calcined petroleum coke) is ground to a specific particle size and blended with a binder of hot pitch to form a semisolid mixture, which is then pressure molded. At some plants this process is largely automated, while at others, employees can be exposed to the dust generated in the grinding and mixing operations and to pitch volatiles. Exposure to oil mist from lubricants may occur during the anode-block pressing operation.

The green-carbon anode is packed in ground coke and baked for several days in furnaces at approximately 1100°C. The volatiles are either contained under pressure in the ovens where they are burned as fuel, or removed and passed to scrubbers. Employees working here may be exposed to pitch volatiles and particles.

After baking, these carbon anode-blocks are moved to the anode rodding section, where they are fitted on rods. The rods are mated with the anode using cast iron poured into a hole in the anode.

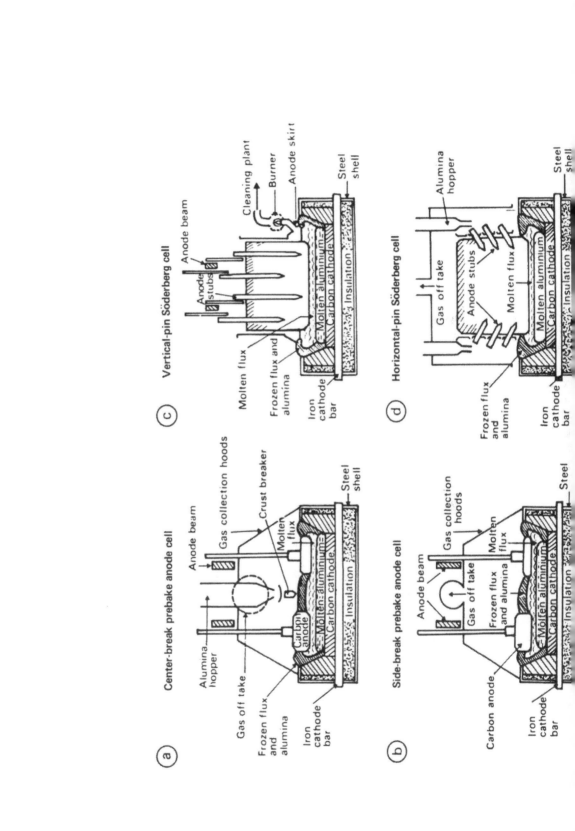

(a) Center-break prebake anode cell

Anode beam
Alumina hopper
Gas off take
Gas collection hoods
Crust breaker
Molten flux
Frozen flux and alumina
Iron cathode bar
Carbon anode
Molten aluminum
Carbon cathode
Insulation
Steel shell

(b) Side-break prebake anode cell

Anode beam
Gas collection hoods
Gas off take
Molten flux
Carbon anode
Frozen flux and alumina
Iron cathode bar
Molten aluminum
Carbon cathode
Insulation
Steel

(c) Vertical-pin Söderberg cell

Anode beam
Anode stubs
Cleaning plant
Burner
Anode skirt
Molten flux
Frozen flux and alumina
Iron cathode bar
Molten aluminum
Carbon cathode
Insulation
Steel shell

(d) Horizontal-pin Söderberg cell

Alumina hopper
Gas off take
Anode stubs
Molten flux
Frozen flux and alumina
Iron cathode bar
Molten aluminum
Carbon cathode
Insulation
Steel shell

There are two designs of prebake cells. In the center-break cells (see Figure 2a), the alumina is fed into the cell either continously or at intervals of 3 to 6 hr. A hard crust, which forms on the melt, is broken in the middle of the cell, and the hood need not be opened. In the side-break type (Figure 2b), the alumina is added between the anodes and the wall of the cell, and the hood must be opened for crust breaking.

When the carbon anodes in a prebake pot are almost completely burned out, they are replaced by a new anode assembly which lasts 20 to 30 days. The work of carbon setters and pot operators may vary from plant to plant depending on the extent to which the process is automated, and can include the manual lowering of the anodes, adjustment of pot voltage, and addition of alumina and fluoride compounds as required. Molten aluminum is siphoned from the pot under vacuum into a large crucible and transported by overhead cranes to holding furnaces in the casting area.

b. Söderberg Anode Process

Typical Söderberg anode cells are shown in Figures 2c and 2d. These differ from the prebake cells in that the anode is baked *in situ*.

The crushing, sizing, weighing, and mixing of the carbonaceous raw materials are done in a manner similar to that used in prebaked anode production. However, the proportion and softening point of the pitch used, and the temperature at which the mixers are operated are different. The warm paste discharged from the mixers may be taken directly to the Söderberg potrooms, or may be formed to briquettes, cooled, and distributed to the pots when needed.

The paste is applied on the top of an anode casing above the pot, and the anode is baked by the heat produced by the passage of the electric current. In one design, pins are inserted vertically into the anode and the current enters by these pins. These pots are known as vertical-pin Söderberg pots. The gases pass to the side of the anode under a skirt and are burned off (see Figure 2c). It has been estimated that in the best case, 5%, and in the worst cases, 40%, of the fumes pass into the potroom.

In another design (horizontal-pin Söderberg cells) the pins in the anode enter at the side (see Figure 2d). The fumes are collected by a hood over the whole cell and collection efficiencies of about 95% are obtainable.

Duties of the pot operators are similar to those in prebaked anode plants. As the anode is consumed and lowered, the pins embedded in the anode must be removed and relocated before they make contact with the molten bath. The vertical-pin design requires that workers stand on a catwalk on top of the anode while adjusting bolts which attach the pins to the overhead buss bar. For pots of the horizontal-pin design, workers operate at the floor level using dollies that are equipped to insert and remove the studs.

When the cathodes in the various cell types have served their useful lives (approximately 3 to 5 years), the pots are relined. The old cathode and crust of solidified bath are removed using pneumatic hammers. The new cathode is made using a paste which is similar to that used for the anode and is made in a variety of shapes to line the bottom of the pots. The cathode may be rebuilt *in situ*, or the old material may be removed and the pot relined in a special work area.

Exposures during rebuilding of the cathode depend on the method used. When the cathode is reconstructed using a ramming paste (a mixture of sized anthracite and about 13% pitch), the mixture is heated to approximately 135°C and rammed or vibrated into place in the pot. The cathode is converted to carbon in the potroom by putting it into the circuit, resulting in the emission of volatiles. When prebaked carbon blocks are used, they are joined together with a warm, liquid pitch paste, resulting in some exposure of workers to pitch volatiles. In some instances, solvents rather than heat are used to make the paste workable.

The concentration of PAH and other contaminants in the potroom atmosphere may be influenced by the type and design of the pot, hooding and hood exhaust rate, building ventilation, size of operation, and electrical current used. The actual exposure of a worker also depends on work practices.[9] In recent years, much has been and is still being done to reduce exposure to PAH through design of new equipment (e.g., crust breakers) and by use of respirators, air-supplied hoods, etc.

C. Iron and Steel Works[3]

1. Iron Production

By far the most important method for the manufacture of iron is the blast furnace process. It involves a large chemical reactor into which iron ore, coke, and limestone are charged. The iron ore usually contains from 50 to 65% iron in the form of oxides. The function of coke is to produce heat and reducing gases when burned with preheated air near the bottom of the furnace. Limestone is added to adjust the composition of the slag (e.g., silica, alumina).

The raw materials are charged continually into the top of the furnace. As the iron ore descends through the furnace, it is reduced to iron which is melted by a countercurrent flow of hot, reducing gas. The molten iron and slag are removed from the hearth of the furnace about every 2 to 4 hr.

The main product from the blast furnace is hot metal or pig iron. This is generally refined to make steel, or it may be used to make iron castings.

2. Steel Production

Steel is a generic name for a group of ferrous metals composed principally of iron. By the proper choice of carbon content and alloying elements, a wide range of mechanical and physical properties of steel may be obtained. The usefulness of steel is based on its abundance, durability, versatility, and low cost.

Steel is manufactured from pig iron or iron and steel scrap. Pig iron consists of iron combined with numerous other elements, including most often carbon, manganese, phosphorous, sulfur, and silicon. The majority of these elements are removed by oxidation. Furthermore, acidic or basic additives are used in the processes to remove the unwanted elements.

There are several different processes used in the production of steel. Until about 1970, the principal method was the open-hearth process. In this process the metallic charge of steel scrap and liquid blast-furnace iron is placed on the bottom of an elongated tunnel-like furnace. A fuel such as heavy fuel oil or tar is heated and injected into the furnace through an end wall. Preheated air burns the fuel, thereby heating the charge until it melts. The oxygen for the oxidation comes from the air above the bath of liquid steel, iron ore added to the charge, and oxygen gas blown down into the liquid steel.

Today the most widely used processes for making steel from molten pig iron are the oxygen steelmaking processes. In these processes, 99.5% pure oxygen is mixed with hot metal, causing the oxidation of excess carbon, silicon, and other elements in the hot metal, and thereby producing steel.

The top-blown basic oxygen process is conducted in a cylindrical furnace. A jet of gaseous oxygen is blown at high velocity onto the surface of a bath of molten pig iron at the bottom of the furnace by a vertical pipe or lance inserted through the mouth of the vessel. Due to intimate mixing, the oxidizing reactions take place very rapidly. The reactions are exothermic and temperature is controlled by adding solid steel scrap or other cooling agents.

In the bottom-blown basic oxygen process, oxygen is introduced into the furnace through pipes in the bottom of the furnace similar to the Bessemer converter. Considerable heat is generated when the oxygen gas oxidizes the carbon and silicon in the molten bath. In order to cool the oxygen inlet, a cylindrical stream of hydrocarbons such as methane surrounds the oxygen jet stream.

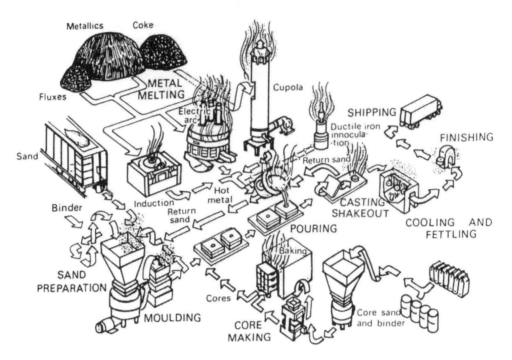

FIGURE 3.   Iron foundry processes.

The direction in which the oxygen is blown has important effects on the final steel composition and on the amount of iron lost in the slag.

Liquid steel is also made by melting steel scrap in an electric furnace. The primary raw material for the electric-furnace steelmaking processes is steel scrap. There are several different processes, the most common being the electric-arc furnace. The thermal energy required for melting the steel scrap is provided by arcs between the graphite electrodes and the charge.

D. Iron and Steel Founding[10]

Foundries produce shaped castings from remelted metal ingots and scrap. Iron and steel foundries generally comprise the following basic sections (see Figure 3): patternmaking, molding, coremaking, melting and pouring, shake-out, and fettling (cleaning and final dressing).

Silica sands and clays have been the principal molding constituents in which metals have been cast from early history. Today, molding sands fall into two broad categories according to the type of base sand employed. Naturally bonded sands are those in which the refractory grains are associated in their deposits with the clay needed for molding. Such sands often develop sufficient properties with the addition of water alone, but their high clay content may reduce refractoriness and gas permeability. Synthetic sands are based mainly on silica sands containing no binders in the natural state. To obtain the necessary strength properties, binders are added separately.

The function of the binders is to produce cohesion between the refractory grains in the green or dried state. Many substances possess bonding qualities; clays, organic oils and resins, cereals, and water-glass silicate may be used singly or in combination. Most ferrous castings are produced in green sand molds. The term "green sand" implies that the binders include clay. Green sand molds also contain organic additives such as cereals, dextrin, starch, wood flour, pitch, or pulverized coal dust. These agents are present in amounts of up to 6% in the sand and provide a reducing atmosphere inside the mold during casting. As the metal is poured, these materials partially decompose, and hydrogen, carbon monoxide, carbon dioxide, water vapor, and volatile hydrocarbons are driven off into the foundry atmosphere. A graphitic layer of lustrous carbon is formed on the surface of sand grains, preventing sand-metal reactions.

The core binders include oils, urea-formaldehyde resins, phenol-formaldehyde resins, and polyurethane resins. The organic binders decompose with the heat during casting. The total emission of gases and vapor is lower in comparison to those from conventional green sand, but their composition is highly dependent on the type of resin system.

After molding and coremaking operations, the mold is ready for pouring. Melting materials, such as ingots, alloys, and scrap are melted in cupolas, electric-induction furnaces, or electric arc furnaces, usually on a batch basis. The molten metal is transported to the casting area in ladles. Cast irons are poured at about 1400°C and high-alloy steels at somewhat higher temperatures. The organization of pouring is a crucial step in founding, since little time is available for casting operations, and an error in the pouring rate or temperature may deteriorate the mold and casting.

In the shake-out section, the casting is removed from the mold with the aid of pneumatic tools, hammers, vibratory tables, etc. The remaining work in producing the finished casting is carried out in the fettling shop and consists of the cleaning and removal of excessive parts from the casting.

E. Petroleum Processing[11]

Crude oil contains a wide range of hydrocarbons, from light gases to residuum that is too heavy to distill even under vacuum. It also contains some sulfur and nitrogen compounds. The function of petroleum refining is to separate crude oil into fractions that are then processed to meet various product specifications.

The modern refinery is a complex and efficient integration of many separate process units and operations, yielding a multitude of end products. Technological advancement has enabled the utilization of almost every fraction of the crude, resulting in very little waste. Most refining processes can be grouped into one of three classes:

1. Separation, e.g., atmospheric and vacuum distillation to give the proper boiling range
2. Conversion, e.g., cracking to change molecular weight and boiling points
3. Upgrading, e.g., catalytic reforming or hydrotreating to meet certain product-quality specifications

Figure 4 gives a schematic flow diagram of the refinery processing steps.

In refineries, PAH occur either as naturally present constituents of the crude oil (1 to 2 wt%[11]), through conversion of the organic material by high temperature processes, or by incomplete combustion of fuel oils in furnaces. The main potential sources of airborne PAH are refinery furnaces and regeneration of cracking catalysts. It has been estimated that approximately 6 tons of BaP are released annually in the U.S. as a result of combusting the carbon on catalyst surfaces to regenerate the catalyst.[12]

During refining, PAH are generally found to concentrate in the higher boiling distillates and solid residues. Petroleum pitch is heavily enriched in PAH (see Table 1).

F. Coal Conversion

The shortage and increasing price of crude oil has resulted in an increased interest in synthetic fuel production on the basis of coal. "Coal conversion" is a generic term encompassing a great number of specific processes designed to produce gaseous and liquid fuels. Generally, these processes are accompanied by the possible production of PAH.[13]

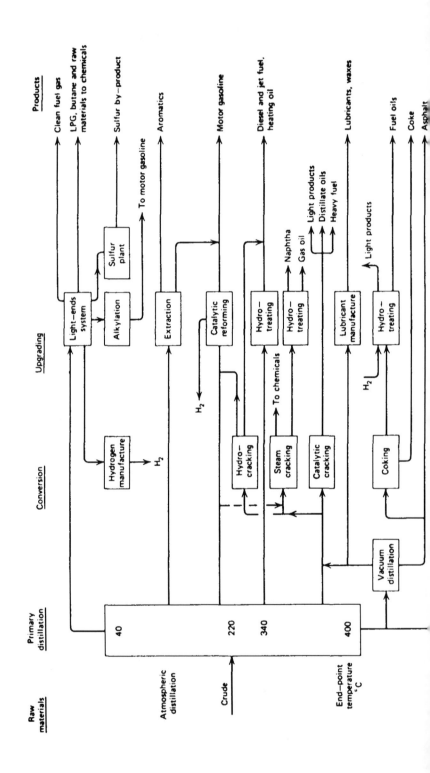

1. Coal Gasification

a. Historical Perspectives

The production of gas from coal was practiced as far back as the early part of the 19th century when coal was destructively distilled in retorts to produce gas for lighting purposes. The gas industry grew rapidly. By the middle of the 19th century, there were few large towns and cities in Europe and North America that did not have gas works and a system for distributing gas.

A variation of coal-derived gas appeared in the late 1800s when water-gas generators were introduced. Air and steam were blown over hot coke in a cyclic manner to produce carbon monoxide and hydrogen. Although of low calorific value, this mixture, water-gas, was used for heating or power generation. Producer-gas reactors, which appeared at about the same time, also converted essentially all coke to gas.

Until about 1960, large amounts of coal were carbonized for the production of town-gas in countries where natural gas was not available. A refined process of coal gasification used oxygen instead of air and was operated under pressure. The first full-scale Lurgi gasifier was introduced in 1936. The Lurgi gasifiers produced a gas of medium-calorific value, which could be used for town-gas production or as synthesis gas in the chemical industry. After catalytic methanation, the product has a composition similar to natural gas. Currently, there are on the order of 100 Lurgi gasifier and other new gasification processes in commercial operation.

b. Modern Coal Gasification Processes

In modern coal gasification, air or oxygen and steam are reacted with coal under various conditions of temperature, flow, and residence time to partially oxidize the coal and produce combustible gases, in which the major fuel constituents are carbon monoxide, hydrogen, and methane. The coal is completely consumed in modern processes, i.e., no coke is produced.

When an air-steam mixture is used to gasify coal, the produced gas is of low calorific value, containing nitrogen as a major component. This gas is suitable for use as a fuel near its point of generation, but is not economical for long-distance transmission. Gas of medium calorific value is obtained when oxygen and steam are used to gasify coal. This gas can be used either as an energy source or as a synthesis gas for the production of chemicals and liquid fuels.

Gas of medium calorific value can be further processed to produce a synthetic pipeline gas of high calorific value, which contains over 90% methane and is very similar to natural gas. Production of methane requires a ratio of hydrogen to carbon monoxide of 3:1. To obtain this ratio, the gases are shifted by reaction of carbon monoxide and water to produce carbon dioxide and hydrogen (water-gas shift reaction). In the presence of a nickel catalyst, hydrogen is then reacted with carbon monoxide to form methane.

The first stage in total gasification is the destruction of the molecular structure of the coal by heat: the coal swells and becomes plastic, and any volatile compounds formed evaporate.

According to the conditions in the chamber, a proportion of these are swept out unchanged as tar and oils with the process gas. The remainder, which is mainly carbon, is partially oxidized, providing most of the heat by which the reaction continues. The reactions taking place within the chamber are extremely complex.

The by-products generated in various coal-gasification processes are of three types: (1) gases such as hydrogen sulfide, carbonyl sulfide, ammonia, and hydrogen cyanide; (2) hydrocarbon gases, aerosols, and residues, including PAH; and (3) mineral particles and wastes.

The production of tars and gases during gasification is, to a great extent, determined by the conditions and dynamics of the gasification processes.

c. Commercial Gasifiers

Modern coal gasifiers may by assigned to three general types of processes: fixed-bed, fluidized-bed, and entrained-bed. A few examples of gasification systems are described later. However, it is emphasized that there is wide variation in the general design and the specifications of the unit operations both within and among different gasification systems. Selection of a system depends on a number of parameters, including the type and end use of the gas produced, e.g., methane, methanol, power generation, etc.

The Lurgi gasifier is a fixed-bed gasifier, pressurized to about 30 atm and fed with crushed coal, oxygen, and steam. A complete Lurgi coal-gasification system for the production of synthetic pipeline gas comprises the following operations: coal handling and preparation, coal feeding, coal gasification, ash removal, gas quenching and cooling, shift conversion, gas purification, methanation, sulfur removal, gas-liquor separation, phenol and ammonia recovery, and by-product storage and cleanup. Several steps might result in exposure to fugitive emissions containing PAH, e.g., coal feeding, injection, gasification, quenching, and methanation.

Lurgi operation is typical of fixed-bed operations in that coal is fed onto the top of a slowly descending bed, while oxidant gas and steam are introduced in the bottom of the gasifier. At the top of the gasifier coal bed, coal is distilled by rising hot gases from below. This releases tars and hot oils into the product gas stream. Beneath the zone of pyrolysis in the bed, the carbon in the residue is gasified at 650 to 815°C by the hot gases from below. At the bottom of the bed, the remaining carbon is oxidized at 980 to 1370°C, providing the hot gases for the reactions in the higher zones. The fixed-bed gasifiers produce relatively large amounts of tars, resulting in 4 to 40ℓ of tar per ton of coal.

The Winkler fluidized-bed gasifier has seen wide development in the past decades. In the Winkler process, air or oxygen is blown up from the bottom of the coal bed at a velocity great enough to partially "fluidize" the bed of coal. Feed coal is crushed to permit fluidization at reasonable gas velocities. The thorough mixing results in bed temperatures of near 1100°C, and in the gasification of most of the tars distilled from the coal.

In the Koppers-Totzek entrained-bed gasifier, pulverized coal is reacted with co-currently flowing oxygen, and a temperature of about 1900°C is reached. Because of the fine size of the coal particles, as well as the co-current high-temperature flow, essentially complete gasification of coal tars is possible.

d. Experimental Gasifiers

Besides these commercial processes, there are several experimental pilot plants which are used to optimize various operation parameters or to develop new processes. An example is the U.S. Department of Energy Morgantown Energy Technology Center (METC) pilot-scale gasifier utilizing a modified Lurgi process. Its major features are a stirring arm, to permit the use of any caking coal as fuel, and a versatile, gas cleanup system which is used to test various hot or wet cleanup methods.

A fluidized-bed hydro-gasifier (HYGAS) process is used by the U.S. Institute of Gas Technology and the Argonne National Laboratory for technology development and testing. The HYGAS process is three stages, producing gas of high calorific value on a pilot scale. In the top stage of the gasifier, hydro-gasification is carried out using the hot fuel gas generated in the low stages. The heavy hydrocarbons formed in this stage are condensed out for use in a recycled oil, which is used to prepare the coal as a slurry for feeding to the hydro-gasifier.

Other processes currently at the pilot plant stage include CO_2^- Acceptor process by the Consolidation Coal Co., the Synthane process by the Bureau of Mines, and the Bi-Gas process by the Bituminous Coal Research.

2. Coal Liquefaction

Coal liquefaction has the advantage over gasification that a greater range of liquid products can be obtained by varying the types of processes and operating conditions. In general, the hydrogen to carbon ratio of liquid hydrocarbon fuels is higher than that of coal for liquefaction. Today there are several major technologies:

1. Pyrolysis
2. Direct hydrogenation
3. Solvent extraction
4. Indirect liquefaction

At present the only commercial operation for production of synthetic liquid fuels form coal is the South African Coal, Gas, and Oil Co. (SASOL) plant in South Africa, which uses indirect liquefaction by the Fischer-Tropsch process. However, there are many experimental pilot plants which are being used in the development of various processes.

As was mentioned earlier, evolution of gaseous and liquid products occurs during destructive distillation of coal. Coal pyrolysis, as a means to generate liquids, has thus far proved to have limited commercial value. However, several modern coal-conversion processes enhance the recovery of potential fuel liquids, especially through the use of catalysts.

Direct hydrogenation of coal involves the addition of hydrogen under pressure to a coal-oil slurry in a catalytic reactor. Examples of recently developed processes are the Synthoil and the H-Coal process.

For solvent extraction, crushed coal is dissolved in a process-derived solvent through an indirect transfer of hydrogen to coal. In the Solvent-Refined Coal (SRC) process, the pulverized coal is mixed with a solvent such as anthracene oil and hydrogen, pressurized, and heated to about 400 to 500°C. The reaction is complete in approximately 20 min, during which time nearly all of the organic matter in the coal is dissolved. The liquid is filtered to remove mineral residues and then fractionated to recover the solvent.

The Exxon donor solvent process is a two step process in which recycled solvent is hydrogenated before it is fed to the liquefaction reactor to transfer hydrogen to the coal. The indirect liquefaction (Fischer-Tropsch process) involves the initial gasification of coal to produce a mixture of CO and H_2 which is purified and then catalytically converted to liquid hydrocarbon fuel. The product distribution obtained varies considerably with temperature, pressure, and catalyst composition. This process is operated in South Africa by SASOL.

Methanol manufacture from CO and H_2 presents an inexpensive alternative for the manufacture of a clean liquid fuel from coal. Methanol may be used as such or converted to hydrocarbon. Recently, the Mobil Corporation developed a process for almost quantitative conversion of methanol to gasoline and water by the use of zeolites. The product is chemically similar to petroleum-derived gasoline.

G. Shale Oil Production[14]

Oil shale consists of sedimentary inorganic material that contains complex organic polymers which are high molecular weight solids (kerogen). It was formed in ancient lakes and seas by the slow deposition of organic and inorganic remains from the bodies of water. Oil shale deposits occur widely throughout the world. The composition of inorganic and organic components of the oil shales varies greatly with deposit location. Certain shales are sufficiently rich in kerogen to allow the production of petroleum substitutes by thermal decomposition (retorting).

The amounts of oil, gas, and coke produced depend on the pyrolysis conditions. Most oil shale retorting processes are carried out at approximately 480°C to maximize liquid-product yield. Retorting processes can be classified as (1) above ground retorts, in which the oil shale mined and transported to metal retorting vessels located on the ground surface, and (2) *in situ* retorts, in which the oil shale is blasted by explosives to form retorts in the ground.

The heat required for above-ground retorting is transferred to the oil shale in four main types of retorts:

1.   External heating by transferring heat through the vessel walls to the oil shale
2.   Internal generation of heat by partial combustion of oil shale
3.   Introduction of externally heated hot gases
4.   Use of externally heated solids (e.g., ceramic balls) to heat the oil shale

At present, only two countries produce oil industrially: the People's Republic of China and the U.S.S.R. However, several other countries, i.e., the U.S., Australia, and Brazil may have oil shale development in the future. A great number of demonstration facilities are in operation for the optimization and development of retorting processes.

The composition of shale oil depends on the shale from which it was obtained as well as on the retorting method by which it was produced. Crude shale oil has a high content of organic nitrogen (ca. 2 wt%) which acts as a catalyst poison and has to be removed prior to upgrading in petroleum refineries. Crude shale oil contains about 3 ppm of BaP, which is about the same as found in petroleum crude oils (see Table 1).

H. Asphalt Production and Use

Asphalt is a black, cementitious, thermoplastic material in which the predominating constituents are bitumens that occur in nature or are obtained in petroleum processing. Since the early 1900s, most asphalts have been produced from the refining of petroleum and used primarily in paving and roofing applications.

Usually, asphalt is obtained by a two stage distillation of crude oil (see Figure 4). The residuum obtained from the vacuum distillation unit is called straight-reduced asphalt. It is a mixture of the very low- and nonvolatile components of the crude oil which are not changed in chemical nature by the process.

Asphalt forms colloidal suspensions consisting of three main components: asphaltenes, resins, and oils. A fractionation scheme for asphalt analysis is given in Figure 5.[15] A typical composition is given in Table 2.

Asphaltenes are polar, polynuclear compounds with alkyl side chains which form large aggregates. Resins are also characterized by aromatic structures; however, they have a lower molecular weight and contain fewer heteroatoms than the asphaltenes. Asphaltenes and resins are dispersed in the oil fraction, which consists mainly of aliphatic and aromatic hydrocarbons.

A process often applied to straight-reduced asphalt is air-blowing. By contact with air at 200 to 275°C the asphalt stock is converted to a harder product. It has been shown that dehydrogenation and polymerization are involved. The oxygen in the air combines with the hydrogen in the asphalt to evolve water vapor leaving unsaturation for cross-linking. Generally, air-blowing forms asphaltenes at the expense of the resin fraction.

Air-blowing may be conducted in batches or in a continuous process, the latter being much more economic with respect to labor requirements and time. Practical use of continuous air-blowing has been made in the manufacture of paving binders from soft vacuum residua, and in the manufacture of roofing asphalt. Figure 6 illustrates the flow of a continuous air-blowing unit.

Due to their waterproof and weather-resistant properties, asphalts have a wide use as protective films, adhesives, and binders. Their largest use is pavements where they serve primarily as binders in paving mixes and as bases in liquid asphalts. The air-blown product is typical of a material used as a steep-pitch roofing asphalt.

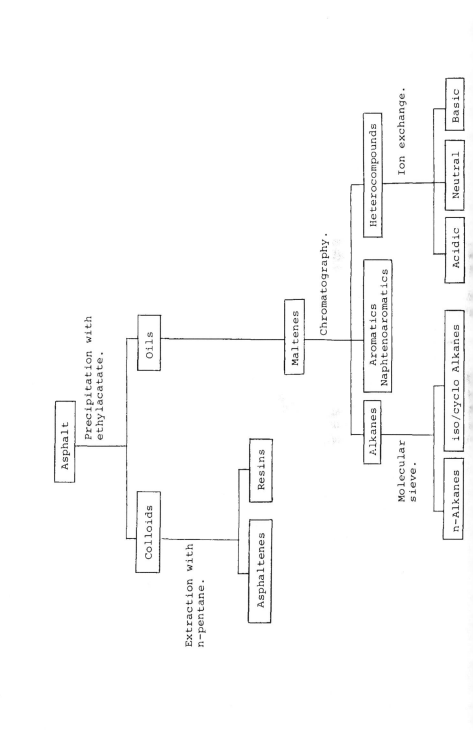

Table 2
TYPICAL COMPOSITION OF
ASPHALT FROM MIDDLE EAST
CRUDE OIL

Compound class	wt%
n-Alkanes	<0.1
Iso/cyclo alkanes	16.1
Aromatics + naphthenoaromatics	38.3
Asphaltenes	15.6
Heterocompounds, acidic	7.2
Heterocompounds, basic	8.5
Heterocompounds, neutral	14.2

From Neumann, H. J., *Erdöl Kohle*, 34, 336, 1981. With permission.

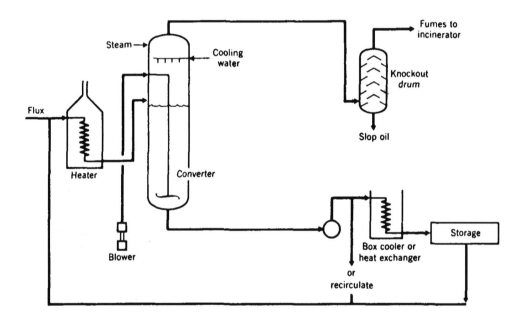

FIGURE 6.   Continuous air-blowing process for asphalt. (From Dickinson, P. F., *Kirk-Othmer Encyclopedia of Chemical Technology*, Vol. 16, 3rd ed., John Wiley & Sons, New York, 1982, 333. With permission.)

Generally, the PAH content of asphalt is low, as is seen in Table 3.[16] The high-boiling petroleum distillation residue probably indicates the upper limit of PAH in asphalt. Frequently, however, asphalt is mixed with coal tar pitch containing much higher amounts of PAH (Table 1). Since the product is handled at elevated temperatures, PAH may evaporate from such mixtures.

I. Carbon Black Production[17]

Carbon blacks are powdered forms of elemental carbon manufactured by the controlled-vapor phase pyrolysis of hydrocarbons. They should not be confused with soot, which is formed by the incomplete combustion of any kind of carbon-containing material.

Carbon blacks have small particle sizes, high surface area per mass unit, quite low content of ash and toluene-extractable materials, and a varying degree of particle aggregation. The degree of "structure" is determined by the size and shape of the aggregated particles, the number of particles per aggregate, and their average mass.

Table 3
PAH CONTENT OF COMMERCIAL ASPHALT AND OF A HIGH-BOILING DISTILLATION RESIDUE (IN ppm)

	Commercial asphalt		High-boiling PAH distillation residue
	A	B	
Fluoranthene	0.10	0.18	1.25
Pyrene	0.17	0.80	4.54
Benzo(a)fluorene	0.02	0.10	0.61
Benzo(b)fluorene	0.02	0.09	0.44
Benzo(ghi)fluoranthene	0.11	0.43	2.97
Chrysene	1.64	5.14	18.95
Benz(a)anthracene	0.13	0.86	1.67
Benzo(b)fluoranthene	0.40	1.60	4.07
Benzo(k)fluoranthene	0.43	1.41	2.91
Benzo(e)pyrene	1.62	6.65	10.44
Benzo(a)pyrene	0.30	1.14	1.85
Perylene	0.11	2.29	5.46
Anthanthrene	0.04	0.30	0.31
Benzo(ghi)perylene	1.37	5.50	4.25
Total	6.37	26.40	59.72

From Neumann, H. J. and Kaschani, D. T., *Wasser Luft Betr.*, 21, 648, 1977. With permission.

Typical, commercially available carbon blacks are acetylene black (high purity and structure), channel black (relatively high oxygen content, low degree of structure), furnace black (irregularly shaped aggregates of spherical particles), lampblack (high degree of structure, low surface area), and thermal black (lowest structure and surface area).

Carbon blacks are powerful black pigments and are used as colorants in a wide variety of materials such as printing inks, plastics, and paints. However, the principal industrial use today (about 90%) is based on their ability to reinforce natural and synthetic rubbers. Addition of carbon blacks markedly improves the properties of vulcanized rubber, particularly the resistance to abrasion, tear strength, stiffness, and hardness.

Acetylene black — Acetylene black is produced in a continuous thermal process. The reaction is initiated by burning the acetylene feedstock with a controlled amount of air. When the reaction temperature is sufficiently high (e.g., 800°C) the air supply is shut off, oxidation ceases, and an exothermic self-sustained dissociation to form hydrogen and acetylene black occurs at temperatures up to 1000°C. Because of its high liquid-absorption properties and its thermal and electrical conductivity, acetylene black is used primarily in the construction of dry-cell batteries and in the production of conductive rubber and plastic products.

Channel black — The channel process was the most important method for making carbon blacks in the past. However, rising natural gas prices, smoke pollution, low yield, and the rapid development of the furnace process resulted in a steady decline in the use of this process until it stopped completely in the U.S. in 1976. In West Germany, a plant still produces channel black by an impingement method using coal-tar residues. The molten material is evaporated by a stream of hot coke-oven gas and is heated to about 370°C prior to reaching the burners. Channel black is used for both rubber reinforcment and as a pigment.

Furnace black — The initial furnace process, which was based on partical combustion of natural gas, has eventually been replaced by the oil-furnace process, which now is used to produce about 97% of total world production. A heavy aromatic feedstock from a petroleum refinery or petrochemical operation is injected by atomization into a high-velocity stream

Table 4
PAH IN BENZENE EXTRACTS OF FIVE
FURNACE BLACKS AFTER 250 HR OF
EXTRACTION

PAH	Amount (mg/kg)
Anthanthrene	<0.5—108
Benzo(def)dibenzothiophene + benzo(e)acenaphthylene	<0.5
Benzofluoranthenes (total)	<0.5— 17
Benzo(ghi)fluoranthene	20—161
Benzo(ghi)perylene	23—336
Benzopyrenes (total)	2— 40
Cyclopenta(cd)pyrene	<0.5—264
Coronene and isomers	13—366
Dimethylcyclopentapyrene and/or dimethylbenzofluoranthene	2— 57
Fluoranthene	10—100
Indeno(1,2,3-cd)pyrene	1— 59
Phenanthrene and/or anthracene	<0.5— 5
Pyrene	46—432

Adapted from Locati, G., Frantuzzi, A., Consonni, G., LiGotti, I., and Bonomi, G., *Am. Ind. Hyg. Assoc. J.*, 40, 644, 1979.

of combustion gases produced by the complete burning of an auxiliary fuel with excess air. The feedstock is converted at 1200 to 1700°C to hydrogen and carbon black in high yields. Downstream, the reaction gases are cooled by spraying with water. Approximately 97% of the carbon black used in the rubber industry is produced by the furnace process. Furthermore, furnace black is used in printing inks, plastics, and paints.

Lampblack — Lampblack is made principally by burning aromatic petroleum oils and coal-tar products such as creosote and anthracene oils in open, shallow pans using a restricted air supply at lower temperatures than the other carbon black processes. The lampblack is separated from the tail gas, densified, and pelletized. Lampblack is used primarily in rubber, where it combines low cost with certain desirable performance characteristics.

Thermal black — In the thermal process, a chamber filled with checkered brickwork is heated to about 1300°C by injecting a burning mixture of gas and air. When the required temperature has been reached, the flow of burning gas is stopped and the hydrocarbon feedstock (usually gas) is injected. Contact with the hot bricks causes the feedstock to crack, forming hydrogen and carbon black. This process is run cyclically using two chambers, one being heated while the other is producing carbon black. The primary use of thermal black is in rubber, and to a lesser extent as a colorant in paints and plastics.

Several studies have demonstrated that carbon blacks contain adsorbed PAH and other polycyclic compounds which are extractable with various solvents. The efficiency of the extraction depends on the extraction time and solvent, the type of carbon black, the relationship between sample weight and solvent volume, and the amount of extractable material.

Locati et al.[18] found that a Soxhlet extraction time of 150 hr was necessary to remove 95% of the benzene-extractable matter in five furnace blacks studied. The higher the molecular weight of the adsorbate, the longer the time necessary to obtain extraction. Exhaustive extraction was not obtained before 250 hr. The PAH identified in extracts of five types of furnace blacks used in tire manufacture are shown in Table 4.

Taylor et al.[19] examined the extraction efficiency of three organic solvents for adsorbates on furnace blacks as measured by the percentage of total extractables and BaP extractability.

Toluene was a better extractant than benzene, and both were superior to cyclohexane which could not remove more than 10% of the benzene-extractable BaP.

Some carbon blacks are subjected to after-treatment by various oxidation processes for particular applications such as pigments in photocopy toners. Fitch and Smith[20] found that in these carbon blacks polar-oxygenated PAH are more prevalent and more readily solubilized than PAH. The extracts also contained nitro-PAH, which probably resulted from nitration of adsorbed PAH during oxidative after-treatment with nitric acid. Several other carbon blacks have been shown to contain nitro-PAH.[21-25] After changes in production technique, the nitro-PAH content of carbon blacks could be reduced to under the detection limit.[21]

J. Combustion Engines

Combustion of gasoline or diesel fuel in motor vehicles has been recognized as one of the major sources of ambient air pollution by PAH.[12,26] During the combustion process, the hydrocarbons in gasoline and diesel fuel react partially with oxygen that is introduced to produce hundreds of new compounds, some of which reach the exhaust. The PAH content of vehicle exhaust is clearly related to the percentage of aromatic compounds in the fuel. An increase in the aromatic content of gasoline was found to be associated with a linear increase in PAH emission.

Additional important factors that influence PAH emission are the fuel to air ratio used for combustion and the engine temperature.[27] An excess of gasoline in the gasoline-air mixture above the stoichiometric ratio considerably increases the amount of PAH emitted. When starting an engine, the working temperature has not yet been reached and the need for high fuel concentrations is inevitable. Therefore, PAH emissions with respect to the combusted amounts of gasoline is increased many times.

Another factor for variation of PAH emission is the age of the engine. Deteriorating engines may alter their fuel composition and resulting combustion characteristics by drawing lubricating oil into the cylinders.

K. Lubricants

Lubricating oils are intended to reduce the friction between surfaces in relative motion. They may also serve other, secondary purposes, including heat transfer, corrosion protection, etc. Major uses include engine oils, hydraulic oils, and metalworking oils.

Shale oil was originally used as a lubricant for the spindles of the cotton mules in the British cotton industry. In the 1920s, the excessive occurrence of cancer of the scrotum observed in cotton mule spinners was shown to be caused by the contact with the lubricant.[28]

Since the 1930s, nearly all the lubricating oils have been produced by refining distillates obtained from petroleum crudes. The trend has been toward highly refined oils with substantially reduced levels of impurities, including PAH. Finished lubricant base oils are predominantly hydrocarbons, but may also contain organic sulfur, oxygen, and nitrogen compounds depending on the refining process. The hydrocarbons are normally a complex mixture of aromatics, naphthenes (cycloparaffins), and paraffins having carbon numbers of 15 or more. The proportions of the different species are responsible for the different characteristics of the base oil.

Generally, the PAH content in modern lubricating oils is quite low, as is shown in Table 5, for unused automotive engine oil. However, as the table reveals, the PAH content may increase strongly during use.[29] The extent of the increase appears to depend on the type of application: up to 10-fold for cutting oils and diesel engine oils, but perhaps 100-fold for gasoline engine oils and heat treating oils.[17] Much of the increase in PAH in engine oil appears to arise from gasoline combustion.[30]

Table 5

**MOST FREQUENT CONCENTRATIONS OF PAH FOUND IN
FRESH AND USED MOTOR OILS(mg/kg)**

PAH	Fresh motor oil (N-22)	Used motor oils from vehicles driven with:	
		Gasoline (N-22)	Diesel (N-10)
Fluoranthene	0.07	50.0	16.0
Pyrene	0.30	135.0	25.0
Chrysene + triphenylene	0.70	40.0	10.0
Benzo(b + j + k)fluoranthene	0.08	30.0	6.0
Benzo(e)pyrene	0.2	35.0	3.0
Benzo(a)pyrene	0.06	23.0	5.0
Perylene	0.06	6.0	0.7
Indeno(1,2,3-cd)pyrene	0.001	1.0	2.0
Benzo(ghi)perylene	0.021	60.0	3.0
Anthanthrene	0.01	8.0	1.2
Coronene	0.02	20.0	1.0

From References 21 and 22.

II. PATHWAYS OF EXPOSURE

Pathways for exposure of workers to PAH are the inhalation of vapors or particles, contamination of skin, and ingestion. The relative importance of each exposure route varies according to the particular technology used or the particular type of work in which the worker is engaged.

Consideration of the health risk posed by the inhalation of vapors of PAH has received little attention. One reason may be that only high-boiling PAH that have correspondingly low vapor pressures have been demonstrated to be carcinogenic (see Table 1, Chapter 1). Equilibrium vapor pressures of pure PAH decrease rapidly as the aromatic rings increase in number. While the vapor concentration at room temperature for naphthalene is 4.6×10^5 $\mu g/m^3$, it is only 0.085 $\mu g/m^3$ for BaP.[31] In the workplace, these values should be regarded as upper limits. The equilibrium vapor concentrations may be considerably reduced by effects of condensation on airborne particulate matter.

Even though exposure to PAH vapors may not pose a direct carcinogenic risk, some of the vapors could exert promotional or co-carcinogenic responses when inhaled together with other vapors or particulate PAH. Perhaps more attention, therefore, should be given to the inhalation of vapors of two-, three-, and even four-ring PAH.

PAH with four or more rings are primarily associated with particles when they occur in the atmosphere. The particles may be deposited in various compartments of the respiratory tract depending on the physical properties of the particles, the air velocity in the respiratory tract, and the branching of the respiratory tract.[32] Particles exceeding 10 μm in aerodynamic diameter usually remain in the upper respiratory tract and do not penetrate into the lungs.

Inhalation of PAH adsorbed on particulate matter may be of principal concern at many workplaces such as coke oven batteries and potrooms of aluminum smelters.

In other industries, skin contact with PAH-containing material may be the major route of human exposure to PAH. This may be true for plants with open-air environments and tight sealing of equipment such as in refineries or coal-conversion plants. Direct contamination of employees skin and clothing during maintenance and repair operations has been recognized as a serious problem.

It is not only at the site of skin contamination that damage by PAH can occur. As PAH are highly lipophilic, they readily penetrate the skin into cells and may cause systemic effects. Animal experiments have shown that PAH metabolites are excreted in urine after percutaneous penetration.[33] At low dose levels, urinary excretion was even higher after percutaneous treatment than after oral administration.

Direct ingestion of particles will be prohibited by good occupational hygiene and will seldom occur. However, the inhalation and swallowing of coarse particles may represent an indirect route of ingestion. The inhaled larger particles will be trapped in the mucous fluids of the upper and middle respiratory system. After transport to the throat, they can be swallowed.

III. SUMMARY AND CONCLUSION

The most important sources of PAH in the work atmospheres are coal tar products, which are produced from the destructive distillation of bituminous coal in gas works and coke plants. Coal tar products are used at elevated temperatures in many industrial processes, such as aluminum smelting, iron and steel production, and foundry and asphalt production, thereby emitting PAH into the work atmosphere.

Coal and shale conversion may constitute new, important sources of PAH. However, until now there has been little information available on the quality and quantity of PAH emitted into the work environment during these processes. In other processes, PAH are generated from incomplete combustion of fuels. Work atmospheres contaminated by motor-exhaust gases generally have a low PAH content.

In addition to inhalation, skin contact may be an important pathway for exposure to PAH.

REFERENCES

1. National Institute for Occupational Safety and Health, Criteria for a Recommended Standard.....Occupational Exposure to Coal Tar Products, Publ. No. 78-107, National Institute for Occupational Safety and Health, U.S. Departments of Health, Education and Welfare, Washington, D. C., 1977.
2. **Kipling, M. D.,** Soots, tars and oils as causes of occupational cancer, in *Chemical Carcinogens,* Searle, C. E., Ed., American Chemical Society, Washington, D. C., 1976, 315.
3. **McGannon, H. E., Ed.,** The Making, Shaping and Treating of Steel, U.S. Steel Corp., Pittsburgh, 1971.
4. U.S. Occupational Safety and Health Administration, Occupational safety and health standards, exposure to coke oven emissions, *Fed. Regist.,* 41, 46, 742, 1976.
5. **Eisenhut, W., Langer, E., and Meyer, C.,** Determination of PAH pollution at coke works, in *Polynuclear Aromatic Hydrocarbons: Physical and Biological Chemistry,* Cooke, M., Dennis, A. J., and Fisher, G. L., Eds., Battelle Press, Columbus, Ohio, 1982, 255.
6. **Bjørseth, A., Bjørseth, O., and Fjeldstad, P. E.,** Polycyclic aromatic hydrocarbons in the work atmosphere. II. Determination in a coke plant, *Scand. J. Work Environ. Health,* 4, 224, 1978.
7. **Hughes, J. P., Ed.** *Health Protection in Primary Aluminum Production, Proc. of a Seminar,* International Primary Aluminium Institute, London, 1977.
8. **Pearson, T. G.,** The Chemical Background of the Aluminum Industry, Monogr. No. 3, The Royal Institute of Chemistry, London, 1955.
9. **Bjørseth, A., Bjørseth, O., and Fjeldstad, P. E.,** Polycyclic aromatic hydrocarbons in the work atmosphere. I. Determination in an aluminum plant, *Scand. J. Work Environ. Health,* 4, 21, 1978.
10. **Beeley, P. R., Ed.,** *Foundry Technology,* Butterworths, London, 1972.
11. **Jahnig, C. E.,** Refinery processes, survey, in *Kirk-Othmer Encyclopedia of Chemical Technology,* Vol. 17, 3rd ed., Grayson, M. and Eckroth, D., Eds., John Wiley & Sons, New York, 1982, 183.
12. **Guerin, M. R.,** Energy sources of polycyclic aromatic hydrocarbons, in *Polycyclic Hydrocarbons and Cancer,* Vol. 1, Gelboin, H. V. and Ts'o, P.O.P., Eds., Academic Press, New York, 1978, chap. 1.
13. **Gammage, R. B.,** Polycyclic aromatic hydrocarbons in work atmosphere, in *Handbook of Polycyclic Aromatic Hydrocarbons,* Bjørseth, A., Ed., Marcel Dekker, New York, 1983, chap. 16.

14. **Dickinson, P. F.**, Oil shale, in *Kirk-Othmer Encyclopedia of Chemical Technology*, Vol. 16, 3rd ed., Grayson, M. and Eckroth, D., Eds., John Wiley & Sons, New York, 1982, 333.

15. **Neumann, H. J.**, Bitumen — neue Erkenntnisse über Aufbau und Eigenschaften, *Erdöl Kohle*, 34, 336, 1981.

16. **Neumann, H. J. and Kaschani, D. T.**, Bestimmung und Gehalt von polycyclischen aromatischen Kohlenwasserstoffen in Bitumen, *Wasser Luft Betr.*, 21, 648, 1977.

17. International Agency for Research on Cancer, *IARC Monographs on the Evaluation of the Carcinogenic Risk of Chemicals to Humans*, Vol. 33, Polynuclear Aromatic Compounds, Part 2, Carbon Blacks, Mineral Oils and Some Nitroarenes, IARC, Lyon, France, 1984.

18. **Locati, G., Fantuzzi, A., Consonni, G., LiGotti, I., and Bonomi, G.**, Identification of polycyclic aromatic hydrocarbons in carbon black with reference to carcinogenic risk in tire production, *Am. Ind. Hyg. Assoc. J.*, 40, 644, 1979.

19. **Taylor, G. T., Redington, T. E., Bailey, M. J., Buddingh, F., and Nau, C. A.**, Solvent extracts of carbon black — determination of total extractables and analysis for benzo(a)pyrene, *Am. Ind. Hyg. Assoc. J.*, 41, 819, 1980.

20. **Fitch, W. L. and Smith, D. H.**, Analysis of adsorption properties and adsorbed species on commercial polymeric carbon, *Environ. Sci. Technol.*, 13, 341, 1979.

21. **Rosenkranz, H. S., McCoy, E. C., Sanders, D. R., Butler, M., Kiriazides, D. K., and Mermelstein, R.**, Nitropyrenes: isolation, identification, and reduction of mutagenic impurities in carbon black and toners, *Science*, 209, 1039, 1980.

22. **Sanders, D. R.**, Nitropyrenes: the isolation of trace mutagenic impurities from the toluene extract of an aftertreated carbon black, in *Chemical Analysis and Biological Tate: Polynuclear Aromatic Hydrocarbons*, Cooke, M. and Dennis, A. J., Eds., Battelle Press, Columbus, Ohio, 1981, 145.

23. **Ramdahl, T., Kveseth, K., and Becher, G.**, Analysis of nitrated polycyclic aromatic hydrocarbons by glass capillary gas chromatography using different detectors, *High Resol. Chromatogr. Chromatogr. Commun.*, 5, 19, 1982.

24. **Ramdahl, T. and Urdal, K.**, Determination of nitrated polycyclic aromatic hydrocarbons by fused silica capillary gas chromatography/negative ion chemical ionization mass spectrometry, *Anal. Chem.*, 54, 2256, 1982.

25. **Giammarise, A. T., Evans, D. L., Butler, M. A., Murphy, C. B., Kiriazides, D. K., Marsh, D., and Mermelstein, R.**, Improved methodology for carbon black extraction, in *Polynuclear Aromatic Hydrocarbons: Physical and Biological Chemistry*, Cooke, M., Dennis, A. J., and Fisher, G. L., Eds., Battelle Press, Columbus, Ohio, 1982, 325.

26. **Grimmer, G., Ed.**, *Environmental Carcinogens: Polycyclic Aromatic Hydrocarbons*, CRC Press, Boca Raton, Fla., 1983.

27. **Grimmer, G., Böhnke, H., and Glaser, A.**, Polycyclische aromatische Kohlenwasserstoffe im Abgas von Kraftfahrzeugen, *Erdöl Kohle*, 30, 411, 1977.

28. **Kipling, M. D. and Waldron, H. A.**, Polycyclic aromatic hydrocarbons in mineral oil, tar and pitch, excluding petroleum pitch, *Prev. Med.*, 5, 262, 1976.

29. **Grimmer, G., Naujack, K.-W., Dettbarn, G., Brune, H., Deutsch-Wenzel, R., and Misfeld, J.**, Untersuchungen über die carcinogene Wirkung von gebrauchtem Motorenschmieröl aus Kraftfahrzeugen, *Erdöl Kohle*, 35, 466, 1982.

30. **Grimmer, G., Jacob, J., Naujack, K. W., and Dettbarn, G.**, Profile of the polycyclic aromatic hydrocarbons from used engine oils — inventory by GCGC/MS — PAH in environmental materials. I, *Anal. Chem.*, 309, 13, 1981.

31. **Pupp, C., Lao, R. C., Murray, J. J., and Pottie, R. F.**, Equilibrium vapour concentrations of some polycyclic aromatic hydrocarbons, As_4O_6 and SeO_2 and the collection efficiencies of these air pollutants, *Atmos. Environ.*, 8, 915, 1974.

32. **Pott, F. and Oberdörfer, G.**, Intake and distribution of PAH, in *Environmental Carcinogens: Polycyclic Aromatic Hydrocarbons*, Grimmer, G., Ed., CRC Press, Boca Raton, Fla., 1983, 130.

33. **Jongeneelen, F. J., Leijdekkers, C.-M., and Henderson, P. T.**, Urinary excretion of 3-hydroxybenzo(a)pyrene after percutaneous penetration and oral absorption of benzo(a)pyrene in rats, *Cancer Lett.*, 25, 195, 1984.

Chapter 5

SAMPLING OF PAH IN WORK ATMOSPHERES

I. FILTER SAMPLING

The PAH compounds which have been demonstrated to exhibit carcinogenic effects in experimental animals, and therefore are of major toxicological interest, usually possess four and more rings.[1] As these PAH exist in the atmosphere mainly associated with particles, the efficient collection of suspended matter has been the primary concern in PAH measurements in the workplace. The most commonly used method has been filtering a known volume of air through a porous filter with a lower pore size limit of 0.1 to 1 μm at a regulated flow rate. The U.S. National Institute for Occupational Safety and Health (NIOSH)[2,3] has recommended the collection of particulate PAH on a glass fiber filter followed by a silver membrane. This arrangement has the following advantages:[4]

1. The silver membrane has an exactly defined pore diameter of 0.8 μm.
2. The glass fiber filter is more evenly loaded over its whole area when backed up by a silver membrane.
3. The membrane has an additional sampling capacity.

Others have used glass fiber filters only,[5,6] polyamide membrane filters,[7] or polytetrafluorethylene (PTFE) membrane filters.[8] The collection efficiency of these filters for suspended particulate matter is generally close to 100%.[9] However, due to the chemical reactivity of most PAH, considerable systematic errors may be introduced in the collection of PAH from ambient air. Losses of PAH may occur by the interaction of these compounds with airborne oxidants and other reactive substances.[10]

The effect of filter media on the chemical integrity of the collected particulates has been reported recently.[11] It has been found that extensive oxidation of filter-collected PAH can occur. The extent of degradation varies greatly with the filter type. Silica or glass fiber filters showed the highest catalytic oxidation potential. Hydroxy-PAH and PAH-quinones were tentatively identified as oxidation products; however, large amounts of unidentified highly polar products were formed as well. The recovery of PAH from filters decreases and the yield of oxidation products increases regularly as the volume of air drawn through the filter increases. This is consistent with the mechanism of degradation caused by gaseous oxidation. The use of inert PTFE membrane filters was found to greatly suppress such degradation problems without eliminating them completely.[11]

The mechanism of PAH oxidative reactions on the filter surfaces is not well understood. The surface properties of the filter media, such as specific surface area, water and vapor affinity, and the polar nature of the surface, are likely to play part of the role in filter reactions with PAH. The polar functionalities associated with glass fiber filters also enhance their ability to adsorb or convert reactive gases, such as NO and SO_2, during sampling. These adsorbed species may react further with PAH.

Another serious problem in the classical air-filtration procedures is the loss of PAH due to re-evaporation as air is being drawn through particulate matter already deposited on the filter. Jackson and Cupps[12] were among the first to point out the problem of sublimation of PAH during sampling in workplace atmospheres. In the conventional sampling of aerosols, Cautreels and Van Cauwenberghe[13] were able to show that large quantities of lower PAH (up to the benzofluorenes) are lost due to this evaporation process. At elevated temperatures even BaP is not immune to sublimation. Recently, however, Lao and Thomas[14] have dem-

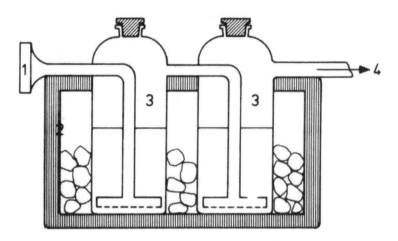

FIGURE 1.   Sampling device for stationary and mobile sampling of PAH.

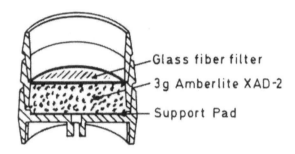

FIGURE 2.   Filter-adsorbent arrangement for personal sampling of particulate and gaseous PAH. (From Andersson, K., Levin, J.-O., and Nilsson, C.-A., *Chemosphere*, 12, 197, 1983. With permission.)

onstrated that particulate PAH collected on glass fiber filters are stabilized by the adsorption to the surface of particulate matter. At high PAH concentrations, when most of the surface sites are covered, the process of evaporation first begins, and loosely bound PAH are released.

II. SAMPLING BY FILTER AND BACK-UP

Evaporative problems can be overcome by trapping volatile PAH in a back-up system. Impingers containing cold ethanol have been used by Bjørseth et al.[5,6] for collecting PAH vapors in workplace atmospheres. The sampling device, which has been used both for stationary and mobile sampling, is shown in Figure 1.

Recently, solid sorbents such as Tenax GC®,[15] Chromosorb®,[12] and Amberlite XAD-2®[8,16] have been demonstrated to efficiently collect organic vapors from air. Several good sampling devices, using both filter and solid adsorbents, have been described. Andersson et al.[16] have developed a simple system for personal sampling consisting of a glass fiber filter and a bed of XAD-2 contained in a commercial plastic filter-holder. A schematic drawing is given in Figure 2. Table 1 shows the distribution of PAH between the filter and the XAD-2 adsorbent from a personal sample taken in the potroom of a Söderberg aluminum plant. A large portion (70%) is found on the adsorbent, including significant amounts of tetracyclic PAH. Even for the benzofluoranthenes, about 3% is found in the adsorbent. This fact could be explained by the rather high air temperature in the vicinity of the pots. At the

Table 1
DISTRIBUTION OF PAH BETWEEN FILTER AND XAD-2 IN A PERSONAL SAMPLE FROM AN ALUMINUM PLANT[a]

Compound	Filter	XAD-2	Total
Naphthalene	<1.0	<1.0	<1.0
Fluorene	0.1	2.8	1.9
Phenanthrene	1.2	26	27
Anthracene	0.13	2.7	2.8
Fluoranthene	1.0	19	20
Benzo(a)anthracene	2.6	1.9	4.5
Chrysene	5.4	2.0	7.4
Benzo(b)fluoranthene	6.2	0.21	6.4
Benzo(k)fluoranthene	4.5	0.13	4.6
BaP	2.4	0.02	2.4
Total	**24.3**	**56.8**	**81**

[a]   Concentrations in $\mu g/m^3$ air.

From Andersson K., Levin, J.-O., and Nilsson, C.-A., *Chemosphere*, 12, 197, 1983. With permission.

Argonne National Laboratory, a device for sampling workplace air at coal-conversion facilities has been developed which consists of a 76-mm PTFE filter, followed by a bed of 120 g of XAD-2 resin.[8] A schematic drawing if given in Figure 3. A principal requirement for sampling in coal-conversion plants is the need for explosion-proof equipment, as often the potential for explosive atmospheres exists in these environments. The sampling system described by Flotard incorporate an explosion-proof vacuum motor with ranges from 0.2 to 36 ℓ/min.

More recent NIOSH-sponsored development of collection systems for PAH has been conducted by Enviro Control. Inc., and has been used to characterize PAH exposure of workers in petroleum refineries.[17] A schematic drawing of the area and the personal-monitoring devices is shown in Figure 4. Both were two-stage units consisting of a silver-membrane filter followed by Chromosorb 102.

III. SAMPLING MODES

Two different modes of sampling are generally used for the measurement of PAH in workplace atmospheres: (1) area sampling using stationary collection devices, and (2) personal sampling using portable collection devices. A third alternative, termed mobile sampling has been used by Bjørseth et al.[6] to determine exposure to PAH at coke ovens. The equipment used was similar to that used for stationary sampling. It was moved around on top of the coke oven battery by the larry car, or carried around on a small backpack. By this procedure, an average of PAH concentrations in various areas was obtained.

A. Stationary Sampling
Stationary sampling using high-volume pumps generally has the advantage that larger quantities of material are available for analysis. Furthermore, the air volume sampled can be determined more accurately than in the case of personal sampling, due to the apparatus used. Stationary sampling has been used for screening purposes, for determining the relative distribution of PAH in particular work environments, or to obtain large amounts of samples for biological testing. In practice, however, the exposure of individual workers at the work-

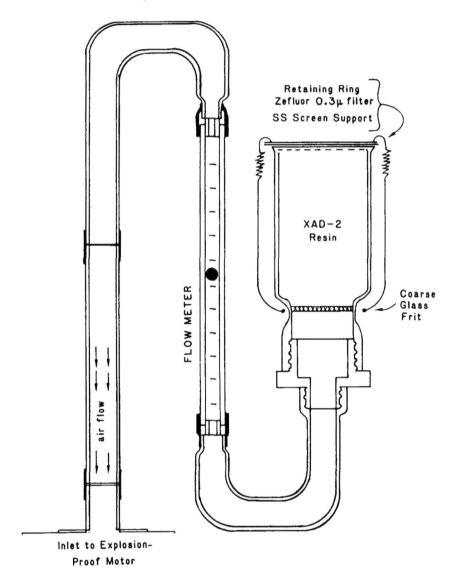

FIGURE 3.   Argonne National Laboratory air sampler with explosion-proof motor. (Courtesy of Argonne National Laboratory, Argonne, Ill.)

place is monitored unsatisfactorily by stationary sampling. First, the PAH do not spread evenly from the source, giving rise to areal variations in airborne PAH, and second, the location of a worker within a workplace may vary considerably during a shift.

B. Personal Sampling

Using personal sampling, a much better picture of the actual exposure of a worker is obtained. Usually, breathing-zone samples are collected by means of portable battery-operated air pumps with filter holders attached to the shirt collars of the workers. In order to satisfy regulatory standards for PAH concentrations, sampling should be performed on a continuous basis through the entire workshift so that integrated samples are obtained, reflecting periods of peak exposure as well as rest breaks. Disadvantages are the generally high error in the determination of sampling volume and the small amount of sample collected due to the low sampling rate (1 to 2 ℓ/min) of the battery-operated pumps. Furthermore,

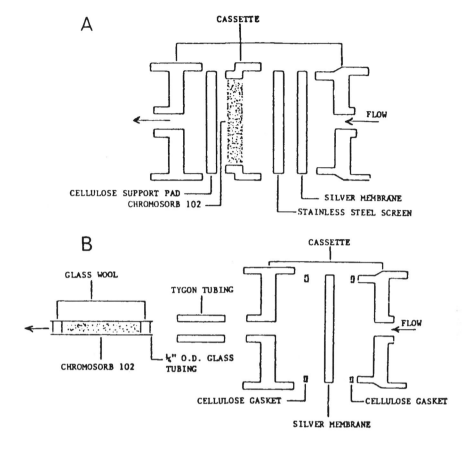

FIGURE 4.   PAH sampling units. (A) For area sampling; (B) for personal sampling.[17] (From NIOSH Technical Report, Petroleum Refinery Workers Exposure to PAHs at Fluid Catalytic Cracker, Coker, and Asphalt Processing Units, U.S. Department of Health and Human Services, Cincinnati, 1983. With permission.)

clogging of filters may occur more easily with the small filters used (usual diameter, 37 mm) in comparison to filters used for stationary sampling.

IV. STORAGE AND SHIPMENT

Due to the reactivity of most PAH, the filters and sorbents should be protected from light and heat after sampling. Photochemical oxidation may lead to considerable losses of PAH deposited on filters when exposed to light.[18] In the absence of light, particulate-adsorbed PAH seem to be quite stable toward oxidation. However, Natusch and co-workers[19] have observed spontaneous dark oxidation of PAH containing benzylic carbon atoms when adsorbed on coal fly ash.

V. SUMMARY AND CONCLUSION

Losses of PAH during filter sampling of particulate matter may occur.

1.   By the escape of volatile PAH and the re-evaporation of already-collected PAH from the filter
2.   Through degradation of reactive PAH under the influence of light or gaseous co-pollutants

Several methods have been described to trap volatile or re-evaporated PAH. The most practicable one seems to be the use of a polymeric sorbent such as Amberlite XAD-2® in a backup section following the filter. Teflon membrane seems to be superior to other filters for the collection of particulate matter, as it minimizes catalytic degradation of PAH during sampling and storage.

After sampling, the filter systems should be stored in the dark and preferably in a refrigerator or freezer to avoid transformation of PAH during storage.

REFERENCES

1. International Agency for Research on Cancer, *IARC Monographs on the Evaluation of the Carcinogenic Risk of Chemicals to Humans*, Polynuclear aromatic compounds. I. Chemical, environmental and experimental data, IARC, Lyon, France, 1983.
2. *NIOSH Manual of Analytical Methods*, Vol. 1. 2nd ed., National Institute of Occupational Safety and Health, Cincinnati, 1977, 217.
3. **Schulte, K. A., Larsen, D. J., Hornung, R. W., and Crable, J. V.**, Analytical Methods used in a Coke-Oven Effluent Study, Publ. No. 74-105, National Institute for Occupational Safety and Health. U.S. Department of Health, Education and Welfare, Cincinnati, 1974.
4. **Steinegger, A. and Glaus, R.**, The determination of polycyclic aromatic hydrocarbons at the workplace in carbon plant and reduction plant, *Light Met.*, 177, 1981.
5. **Bjørseth, A., Bjørseth, O., and Fjeldstad, P. E.**, Polycyclic aromatic hydrocarbons in the work atmosphere. I. Determination in an aluminum reduction plant, *Scand. J. Work Environ. Health*, 4, 212, 1978.
6. **Bjørseth, A., Bjørseth, O., and Fjeldstad, P. E.**, Polycyclic aromatic hydrocarbons in the work atmosphere. II. Determination in a coke plant, *Scand. J. Work Environ. Health*, 4, 224, 1978.
7. **Becher, G., Haugen, Aa., and Bjørseth, A.**, Multimethod determination of occupational exposure to polycyclic aromatic hydrocarbons in an aluminum plant, *Carcinogenesis*, 5, 647, 1984.
8. **Flotard, R. D.**, Sampling and Analysis of Trace-Organic Constituents in Ambient and Workplace Air at Coal-Conversion Facilities, Rep. ANL/PAG-3, Argonne National Laboratory, Argonne, Ill., 1980.
9. **Walter, J. and Reischl, G.**, Measurements of the filtration efficiencies of selected filter types, *Atmos. Environ.*, 12, 2015, 1978.
10. **Nielsen, T., Ramdahl, T., and Bjørseth, A.**, The fate of airborne polycyclic organic matter, *Environ. Health Perspect.*, 47, 103, 1983.
11. **Lee, F. S.-C., Pierson, W. R., and Ezike, J.**, The problem of PAH degradation during filter collection of airborne particulates, in *Polynuclear Aromatic Hydrocarbons: Chemistry and Biological Effects*, Bjørseth, A. and Dennis, A. J., Eds., Battelle Press, Columbus, Ohio, 1980, 543.
12. **Jackson, J. O. and Cupps, J. A.**, Field evaluation and comparison of sampling matrices of polynuclear aromatic hydrocarbons in occupational atmospheres, in *Carcinogenesis: A Comprehensive Survey*, Vol. 3, Jones, P. W. and Freudenthal, R. I., Eds., Raven Press, New York, 1978, 183.
13. **Cautreels, W. and Van Cauwenberghe, K.**, Experiments on the distribution of organic pollutants between airborne particulate matter and the corresponding gas phase, *Atmos. Environ.*, 12, 1133, 1978.
14. **Lao, R. C. and Thomas, R. S.**, The volatility of PAH and possible losses in ambient sampling, in *Polynuclear Aromatic Hydrocarbons: Chemistry and Biological Effects*, Bjørseth, A. and Dennis, A. J., Eds., Battelle Press, Columbus, Ohio, 1980, 829.
15. **Malaiyandi, M., Benedek, A., Halko, A. P., and Bancsi, J. J.**, Measurement of potentially hazardous polynuclear aromatic hydrocarbons from occupational exposure during roofing and paving operations, in *Polynuclear Aromatic Hydrocarbons: Physical and Biological Chemistry*, Cooke, M., Dennis, A. J., and Fisher, G. L., Eds., Battelle Press, Columbus, Ohio, 1982, 471.
16. **Andersson, K., Levin, J.-O., and Nilsson, C.-A.**, Sampling and analysis of particulate and gaseous polycyclic aromatic hydrocarbons from coal tar sources in the working environment, *Chemosphere*, 12, 197, 1983.
17. NIOSH Technical Report Petroleum Refinery Workers Exposure to PAHs at Fluid Catalytic Cracker, Coker, and Asphalt Processing Units, National Institute for Occupational Safety and Health, U.S. Department of Health and Human Services, Cincinnati, 1983.
18. **Peters, J. and Seifert, B.**, Losses of benzo(a)pyrene under the conditions of high-volume sampling, *Atmos. Environ.*, 14, 117, 1980.
19. **Korfmacher, W. A., Natusch, D. F. S., Taylor, D. R., Mamantov, G., and Wehry, E. L.**, Oxidative transformations of polycyclic aromatic hydrocarbons adsorbed on coal fly ash, *Science*, 207, 763, 1980.

Chapter 6

ANALYTICAL METHODS FOR AIRBORNE PAH

There exist numerous methods which may be used for the determination of PAH in work atmospheres. The method of choice depends on our requirements as to degree of detailed knowledge of sample composition, speed, cost, precision, and accuracy. In the following, the relevant methods will be divided in three groups.

1.   Real-time techniques
2.   Fast and inexpensive group- or compound-specific methods
3.   Detailed analysis

I. REAL-TIME TECHNIQUES

The industrial hygienist occasionally needs to quickly evaluate a hazardous situation in terms of the potential human exposure. Therefore, in addition to measurements of integrated exposures obtained by personal monitors, real-time or near-real-time continuous monitors are desirable to warn of elevated concentrations in work areas. Portable real-time monitors will also prove valuable in detecting small leaks in process streams and pinpointing sources of exposures. Several instrumental techniques have been described which may satisfy these needs.

A second-derivative, wavelength-modulated, UV-spectrometer (DUVAS) in portable form (with the added power of a microcomputer to correct for interfering compounds) has been developed at Oak Ridge National Laboratory.[1]

Since a second-derivative spectrum is a measure of the curvature of the absorption spectrum and not of the transmitted light, the DUVAS technique has the important advantage of being independent of sample opacity, light intensity fluctuations, and source-energy variations.[1] This advantage is particularly useful when no sample preparation to remove particulate matter is performed prior to analysis.

The spectrometer is capable of measuring volatile PAH in work atmospheres as demonstrated in Figure 1. The presence of naphthalene and 2-methylnaphthalene is shown in the headspace above a solvent-refined coal (SRC) light oil at ambient temperature. The sensitivity is high, with the detection limit for naphthalene being in the few ppb range (Figure 2).[2]

For rapid screening of liquid or solid samples, a room-temperature phosphorescence (RTP) technique[3] shows promise as a chemical spot test, giving compound-specific information. A liquid sample such as wipe sample dissolved in an organic solvent, is spotted on filter paper. Phosphorescence is selectively induced by treatment with a heavy atom perturber such as lead acetate. It is possible to identify and roughly quantify a major PAH compound without any fractionation. An example is given in Figure 3. Raw Synthoil coal liquifaction tar (15 μg) was dissolved in methylene chloride before spotting on filter paper, The RTP spectrum shown in Figure 3 is specific to pyrene in the amount of 5500 ± 1600 ppm. Independent chromatographic analysis of the Synthoil gave a pyrene concentration of 4.300 ppm.

A passive device capable of measuring integrated exposures to airborne PAH has been developed based on RTP.[3] Filter paper treated with heavy atom salt acts as the collecting element during exposure, and as the substrate from which the RTP can be induced for analysis. Selective PAH such as pyrene in fluoranthene can be measured in concentration x time. The device is calibrated by exposure to the vapors of pure PAH compounds. At equilibrium vapor concentrations, the intensity of the RTP increases linearly with the time of exposure for up to 8 hr.

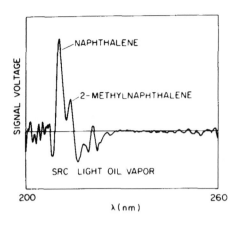

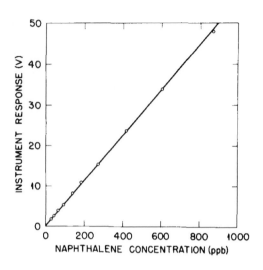

FIGURE 1. Second-derivative UV-absorption spectrum of aromatic headspace vapors above an SRC light oil. (From Gammage, R. B. and Bjørseth, A., *Polynuclear Aromatic Hydrocarbons: Chemistry and Biological Effects*, Bjørseth, A. and Dennis, A. J., Eds., Battelle Press, Columbus, Ohio, 1980, 565. With permission.)

FIGURE 2. Calibration curve for naphthalene vapor. (From Gammage, R. B. and Bjørseth, A., *Polynuclear Aromatic Hydrocarbons: Chemistry and Biological Effects*, Bjørseth, A. and Dennis, A. J., Eds., Battelle Press, Columbus, Ohio, 1980, 565. With permission.)

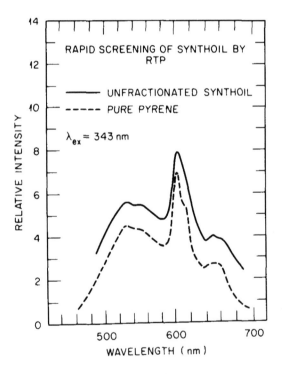

FIGURE 3. Identification of pyrene in unfractionated Synthoil by selective RTP. (From Gammage, R. B. and Bjørseth, A., *Polynuclear Aromatic Hydrocarbons: Chemistry and Biological Effects*, Bjørseth, A. and Dennis, A. J., Eds., Battelle Press, Columbus, Ohio, 1980, 565. With permission.)

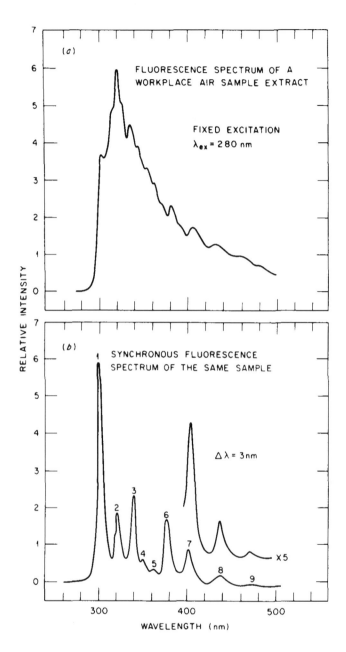

FIGURE 4. (a) Fixed excitation fluorescence spectrum and (b) synchronous fluorescence spectrum of a workplace air sample extract. Peak assignments: 1, fluorene; 2, naphthalene + acenaphthene; 3, benzo(b)fluorene; 4 and 5, chrysene; 6, anthracene; 7, BaP; 8, perylene. (From Vo-Dinh, T., Gammage, R. B., and Martinez, P. R., *Anal. Chem.*, 53, 253, 1981. With permission.)

Complementary information of PAH in complex workplace samples may be obtained by using synchronous luminescence (SL) which involves scanning of both excitation and emission wavelengths.[4] SL results in a bandwith narrowing and a simplification of the spectral profile.[5] This is demonstrated in Figure 4 for a workplace air sample extract.

The ability of RTP and SL techniques to measure major components in unfractionated samples, within a few minutes, opens the door to simple and rapid PAH-indicator monitoring.

Table 1
COMPOSITION OF BENZENE-SOLUBLE ORGANICS OF VARIOUS SAMPLES FROM AN ALUMINUM SMELTER

| Location | Conc. total benzene solubles (mg/m³) | Aromatics (%) | | | | Aliphatics | % Of benzene solubles identified |
		3 Cycles	4 Cycles	5 Cycles	Total		
H. S. Söderberg Line	0.20	7.6	54.2	22.1	84	5	89
Potlining center	0.10	5.7	9.7	1.7	17	83	100
Prebake anode press	0.68	7.3	18.9	0.6	27	90	117[a]
Prebake mixers	0.26	3.6	16.5	5.3	25	4	29
Söderberg mixers	0.53	8.1	34.8	5.1	48	4	52
Ring furnace	0.06	9.6	8.6	1.1	19	73	92

Note: Each value is the average of four ambient air samples taken by high volume sampler. These spot samples do not represent 8 hr employee exposure.

[a] High value may be due to unrepresentative calibration standard.

From MacEachen, W. L., Boden, H., and Larivière, C., *Light Met.*, 2, 509, 1978. With permission.

A much more sophisticated and expensive means of measuring PAH vapors in real-time is available in the form of an atmospheric-pressure chemical-ionization mass spectrometer (APCI/MS).[6-8] The spectrometer can perform essentially instantaneous analyses of air at ambient temperatures. Current detection limits are in the low ppt range for most PAH. Hitherto, this technique has been applied mainly to the analysis of PAH and other combustion products in stack and exhaust gases. However, the mobile spectrometry system will also permit surveys and assessments of PAH vapors to be conducted in workplace atmospheres.

II. FAST AND INEXPENSIVE GROUP- OR COMPOUND-SPECIFIC METHODS

Several methods exist for determination of PAH as a group. The method that probably has been used most is the determination of the benzene soluble fraction (BSF). Following sample collection, the particulate matter captured on filters is ultrasonically extracted with benzene. An aliquot of this extract is evaporated and the residue weighed. From the residue weight and sample volume, the concentration of coal tar pitch volatiles (CTPV) is determined. Although in the U.S. the BSF method is still the only officially accepted method of measuring PAH in the workplace atmosphere, there is a widespread criticism of its accuracy and selectivity.[9,10] As its name implies, it is nonspecific. It analyzes for all benzene-soluble compounds. It has been documented, however, that the working environment contains a significant amount of hydrocarbons that originate from sources other than coal tar pitch, and that are less hazardous. Oils on hot metal parts, lubricants in compressed air, and exhaust from combustion engines may be sources of non-PAH materials soluble in benzene.

A good demonstration of the variation in composition of benzene solubles is given by MacEachen and co-workers[10] who determined the aliphatics and aromatics in samples collected in an aluminum smelter. As seen in Table 1, the percentage of aliphatics varied from 4 to 90, depending on the sampling site.

Figure 5 shows the relationship of PAH to BSF in different locations of aluminum plants.[11] Except for the Söderberg potroom, one finds considerable scattering of the data. There was essentially no correlation between PAH and BSF in the prebake potroom samples, whereas those samples taken in the Söderberg potroom were highly correlated — the correlation

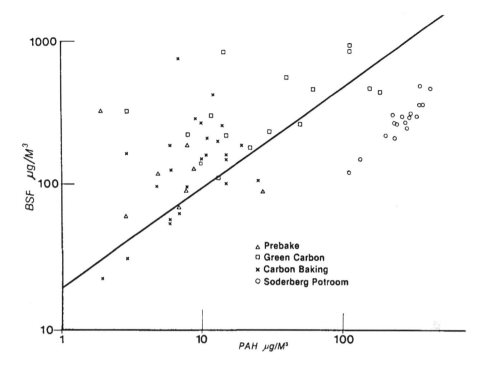

FIGURE 5. Relationship between BSF and the PAH content of particulate matter from different aluminum plant locations. (Courtesy of T. B. Bonney, Aluminum Company of America, Pittsburgh, Pa.)

coefficient being 0.98. Carbon baking and green carbon values showed correction coefficients of 0.51 and 0.68, respectively.

Another problem is the inaccuracy that can occur in the gravimetric determination. The estimated weight of benzene-soluble material can be erroneous due to several factors, including:

1. Disintegration of the filter
2. Loss of particulates during handling and extraction
3. Change in the water content of the filter between measurement and final weight after extraction
4. Loss of PAH during solvent evaporation

Because of the more recent awareness of the carcinogenic nature of benzene, the less toxic solvent cyclohexane has been recommended as a substitute in the extraction of PAH from particulate matter.[12] However, studies have shown that the dissolving capacity of cyclohexane for coal tar pitch is about 25% less than that of benzene.[11] This is observed particularly with the higher molecular weight polycyclic compounds which are of primary toxicological concern.

From the above it is apparent that the gravimetric determination of benzene- (or cyclohexane) soluble fractions is ambiguous as a measure of total particulate PAH and can give no dependable estimate of the degree of hazard. On the other hand, the benzene-soluble fraction itself may subsequently provide a meaningful basis for more definitive analyses.

Many analytical laboratories within the aluminum industry have evaluated various methods to separate the PAH in the BSF from interfering aliphatic material.[13,14] Balya and Danchik[14] have reported on the liquid chromatographic separation of PAH and oil on small, disposable silica columns. The separation is performed by first passing a nonpolar solvent such as cyclohexane through the column and collecting the eluents as "oil" fraction. This is followed

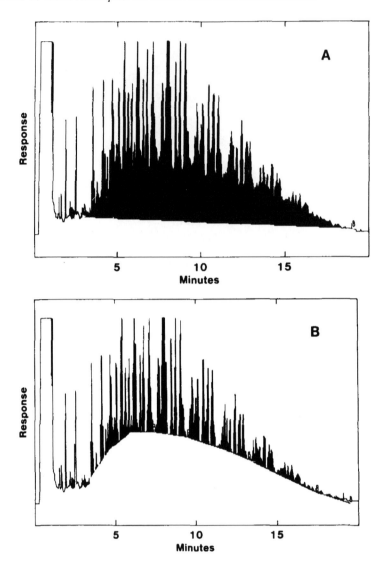

FIGURE 6.   Reconstructed capillary gas chromatogram of green mill exhaust sample using (A) area summation and (B) baseline correction. (From Balya, D. R. and Danchik, R. S., *Am. Ind. Hyg. Assoc. J.*, 45, 260, 1984. With permission.)

by dichloromethane to rapidly elute the PAH fraction. The separation may be viewed under a long-wavelength UV lamp. Another approach proposed to characterize the benzene-soluble matter is based on direct capillary gas chromatographic analysis of the benzene extract.[14] Interfering mineral oil results in an unresolved "hump", while major PAH are easily recognized as distinct peaks. This is demonstrated in Figure 6. The total area (upper chromatogram) is calculated and the unresolved portion removed (lower chromatogram) with the aid of a laboratory data system. This yields the amount of oil and PAH present, provided an appropriate sample of the contaminating oil is available for calibration.

A quick and simple method that has been used over the past few years in several laboratories in Norway is a thin-layer chromatographic (TLC) separation on silica of PAH from other compounds, scraping off the spot for PAH, extracting the PAH from silica, and determining the total PAH content by UV spectrophotometry at 254 nm.[15] The method has to be calibrated against more advanced analytical methods, such as gas chromatography or high-performance

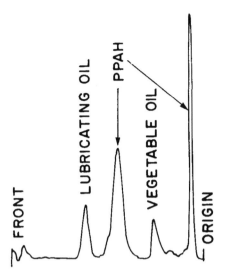

FIGURE 7. Iatroscan chromatogram of an aluminum plant air-monitoring sample. (From Boden, H. and Roussel, R., *Light Met. (Warrendale, Pa.)*, 1049, 1983. With permission.)

liquid chromatography. Provided the same source is studied over a long period of time, the precision is about 15%. The sensitivity is about 5 μg PAH per cubic meter for a sample size of 1 m³ air.[15]

Recently, Boden and Roussel[13] described the use of the Iatroscan technique to separate PAH from interfering compounds in the BSF. In the Iatroscan system, TLC separation is carried out on fused silica rods, 0.9 mm in diameter and 15 cm long, which are covered with an activated silica gel coating. After applying the sample extract close to one end of the rod, separation is achieved by development with a suitable solvent, e.g., 5% diethylether in n-hexane. The rod is dried and scanned in a flame ionization detector (FID). The area under the FID signal is used for quantitation.

Figure 7 shows a typical Iatroscan chromatogram of an air-monitoring sample from an aluminum plant. The separation of polycyclic compounds from interfering substances is clearly demonstrated.

Table 2 gives the results found in a broad survey in an aluminum smelter. There is a very high correlation between the sum of the constituents found by Iatroscan and the gravimetric benzene solubles. However, as shown in the last column of the table, the proportion of particulate PAH in the BSF was found to vary from 10 to 100%. This again indicates that samples of air particulates from workplace environments may contain significant quantities of materials which interfere with the gravimetric BSF method. Thus the Iatroscan technique represents a valuable potential method for the direct measurement of total PAH in samples from workplace atmospheres.

BaP has been widely used as a proxy compound representing the large and complex group of PAH compounds present in air particulate samples. The main reasons for this are that BaP is recognized as one of the most potent carcinogens among PAH, that its occurrence is ubiquitously linked to complex PAH emissions, and that it is relatively easy to quantitate with high sensitivity.

A method for the specific determination of BaP is used in Sweden,[16] where there exists a threshold limit value for BaP. A filter sample is first vacuum sublimed, then the sample is separated by TLC on acetylated cellulose and quantified by fluorescence scanning on the plate. The recovery is determined by an isotope dilution technique to better than 90% at

Table 2
DETERMINATION OF CONSTITUENTS OF BENZENE SOLUBLES
USING THE IATROSCAN TH10

	Iatroscan results (mg/mℓ)				Benzene solubles (mg/mℓ)	% PAH in benzene solubles
Sample identity	Lubricating oil (A)	PAH (B)	Vegetable oil (C)	Constituents (A + B + C)		
HVA-2	—	1.20	–	1.20	1.20	100
-4	—	0.58	—	0.58	0.64	90
HVB-20	0.07	0.13	0.02	0.22	0.21	62
-20	0.03	0.20	—	0.23	0.23	87
HVC-38	0.09	0.13	—	0.22	0.29	45
-39	0.08	0.07	—	0.15	0.18	39
HVD-16	0.01	0.04	<0.01	0.06	0.07	57
-21	0.02	0.05	<0.01	0.07	0.09	56
HVE-1	0.03	0.11	0.01	0.15	0.12	92
-3	0.03	0.14	0.01	0.18	0.16	88
HVF-8	0.01	0.05	0.01	0.06	0.08	63
-10	0.04	0.12	0.04	0.20	0.17	71
HVG-11	0.01	0.15	0.10	0.26	0.22	68
-17	0.02	0.08	0.01	0.12	0.11	73
HVH-25	0.10	0.16	—	0.25	0.26	62
-30	0.06	0.17	—	0.22	0.26	65
HVJ-6	0.11	0.53	0.01	0.65	5.71	80
-7	1.75	1.00	4.00	6.75	0.11	18
HVK-23	0.04	0.05	<.001	0.10	0.18	45
-27	0.04	0.17	—	0.21	0.10	94
HVL-33	0.05	0.06	—	0.10	0.14	60
-34	0.03	0.13	—	0.16	0.22	93
HVM-32	0.07	0.26	—	0.32	0.22	118
-35	0.03	0.16	—	0.18	0.13	123
HVN-12	0.03	0.04	0.01	0.07	0.08	50
-13	0.02	0.02	<0.01	0.04	0.06	33
HVO-13	0.40	0.04	0.12	0.56	0.60	7
-18	0.07	0.03	—	0.10	0.12	25
HVP-22	0.02	0.02	<0.01	0.04	0.05	40
-26	0.03	0.02	—	0.04	0.05	40
HVQ-37	0.08	0.12	—	0.20	0.21	57
-40	0.08	0.19	—	0.17	0.19	47

From Boden, H. and Roussel, R., *Light Met. (Warrendale, Pa.)*, 1049, 1983. With permission.

levels as low as 0.8 pmol. The relative SD was found to vary from 13% for 2.4 pmol, decreasing to 5% for 46 pmol BaP in the sample (seven replicates). The detection limit for authentic samples was found to be about 10 pmol (0.25 ng) per filter. A similar technique using TLC and specific fluorimetric detection has been used by Blome[17] for the determination of BaP in a wide variety of workplace environments.

Other BaP-specific methods are based on high-performance liquid chromatography (HPLC) with selective detection. Boden[18] used both adsorption and reverse-phase HPLC columns in conjunction with a variable UV detector to develop sensitive and selective procedures for the determination of BaP in coal tar pitch volatiles. Tomkins et al.[19] describe the isolation and quantitation of BaP from natural, synthetic, and refined crudes by a sequential HPLC procedure. A BaP-enriched fraction is obtained after normal-phase chromatography on a bonded amino-cyano packing material. The isolated fraction is reinjected onto a reversed-phase column, and BaP is quantified by fluorescence detection.

Low-temperature fluorescence techniques utilizing the Shpol'skii effect have been applied

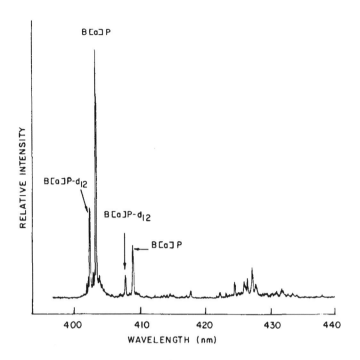

FIGURE 8. Selectively excited fluorescence spectrum of BaP in a shale oil sample with BaP-d_{12} added as internal standard. (From Yang, Y., D'Silva, A. P., and Fassel, V. A., *Anal. Chem.*, 53, 2107, 1981. With permission.)

to the analyses of specific PAH, such as BaP, in a great number of environmental samples.[20] The Shpol'skii effect is operative in frozen solutions of certain solvents, such as n-alkanes, and results in an increase of fluorescent intensity and extreme narrowing of the lines. Further sharpening is possible with laser excitation. Recently, Yang et al.[21,22] have reported on the direct determination of BaP in liquid fuels by laser-excited Shpol'skii spectrometry. The quantitative determination is facilitated by the addition of deuterated BaP as an internal reference.[22] A typical selectively excited Shpol'skii fluorescence spectrum of BaP in a shale oil sample with 10 ppb BaP-d_{12} as internal reference is shown in Figure 8. The simplicity and speed of the procedure, together with its high sensitivity should make it a potential method for the rapid determination of BaP in workplace samples.

Recently, a simple spot test was reported based on sensitized fluorescence.[23] By adding naphthalene to a sample, the fluorescence is significantly increased for most PAH. By comparing spots and nonsensitized samples on a filter paper and subsequently diluting the sample until no difference is observed, an estimate of the PAH content can be obtained. The method was introduced as a qualitative test for environmental samples, and some semiquantitative tests have been made on effluent from woodburning stoves.[23] The method requires no instrumental resources — only a filter paper, a pipette, and a few standard solutions. A test is usually done within 5 to 10 min.

One aluminum plant in Norway has applied this technique, and the results were compared to those obtained by capillary gas chromatography (GC). As shown in Table 3, there is a reasonably good correlation between GC and spot-test results. This indicates that the spot-test may be used at least as a low-cost screening technique to make a first rough estimate of the PAH content of a sample.

III. DETAILED ANALYSIS

Although nonspecific group or proxy methods for PAH may be suitable for screening or

Table 3
COMPARISON OF TOTAL PAH
DETERMINED BY SENSITIZED
FLUORESCENCE AND CAPILLARY GAS
CHROMATOGRAPHY USING SAMPLES
FROM AN ALUMINUM SMELTER (μg)

Sample No.	Fluorescence	Gas chromatography
1	1	0.6
2	1	3
3	10	2
4	10	4
5	10	6
6	10	10
7	10	12
8	10	17
9	100	9

compliance purposes, there are two important facts in favor of a quantitative determination of individual PAH in workplace samples.

First, the biological activity of PAH species may change markedly with relatively minor structural changes, in ways that are difficult to predict. It is, for example well established that of the two isomers, BaP and BeP, only BaP is a proven carcinogen in experimental animals.[24] In addition, substituent groups such as methyl groups, in some positions on the rings of certain PAH, can increase the carcinogenic activity.[25] While chrysene is only a weak carcinogen, 5-methylchrysene has been demonstrated to be highly carcinogenic in multiple species.[24] On the other hand, substitution of a methyl group may also decrease the activity of a highly carcinogenic PAH. 5-methyl- and 5,8-dimethyldibenzo(a,i)pyrene are much less carcinogenic than the parent PAH.[26] Therefore, for a quantitative assessment of the hazard connected with exposure to a complex mixture of PAH, determination of individual components is necessary.

Second, often the relative distribution of individual PAH in a sample, i.e., the PAH profile, is characteristic for the source, or a particular work environment. Therefore, the detailed analysis of individual constituents improves our ability to describe a given workplace, and is a prerequisite for the selection of simplified PAH-monitoring methods.[27]

Most methods for a detailed PAH analysis described in the literature are based on three distinct steps:

1. Desorption of the collected compounds from the sampling matrix
2. Isolation of the PAH from other co-extracted compounds
3. Separation of the PAH mixture into individual compounds and subsequent identification and quantitation

A. Desorption

The objective of the desorption procedure is to separate PAH from the bulk-sample matrix in as high a yield as possible, with a minimum of co-extraction of other compounds present in the sample. The various methods available for desorption of PAH from filters or adsorption traps may be divided into solvent extraction and thermal desorption.

The principle behind solvent extraction is to selectively dissolve the sampling matrix with a suitable solvent. Soxhlet extraction has for many years been a standard method for preparing PAH extracts.[28,29] However, some major deficiencies have been reported in the literature: time-consuming procedures, thermal decomposition of PAH during extraction, and low recovery at low PAH concentrations.[29]

It is possible to reduce extraction time, and to improve extraction efficiency and reproducibility by use of ultrasonic agitation.[30,31] This procedure differs from other solvent-extraction methods in its use of high-intensity ultrasonic vibration (ca. 20 kHz) to produce solvent cavitation around the sample particles, presumably leading to enhanced solvent contact and mixing with the sample. Ultrasonic extraction is included in several standard procedures. It is specified in a NIOSH procedure for determining air particulate BaP[32] and total air particulate aromatic hydrocarbons,[33] and in the procedure proposed by the U.S. Intersociety Committee on Recommended Methods for total air particulate hydrocarbons.[34]

A wide variety of solvents have been used for solvent extraction of airborne PAH. The most common are acetone, benzene, cyclohexane, methanol, dichloromethane, diethylether, carbon disulfide, and tetrahydrofuran. Benzene has been recommended in preference to several other solvents in extracting organic compounds from atmospheric particulate matter due to its high extraction efficiency and selectivity for polycyclic aromatic compounds.[35] However, since cyclohexane extracts fewer extraneous materials,[36] such as asphaltic tars and other uncharacteristic materials, and is less hazardous, it has been widely used as a solvent for PAH extraction.[37-41] Cyclohexane has been shown to be as efficient in Soxhlet extraction of BaP from airborne particulate matter as benzene,[42] while other studies suggest that cyclohexane is less efficient for the extraction of coal tar pitch volatiles, particularly for the higher molecular weight compounds.[11]

Dichloromethane has been found to be superior to benzene for the extraction of organic material from airborne particles.[29,43] An advantage of this solvent might be its lower bp compared to benzene and cyclohexane, reducing the losses during the solvent concentration step. Extensive losses of the more volatile PAH have been reported during the evaporation of higher boiling solvents, particularly during the final stages of concentration.[44] As much as 80% of the diaromatics can be lost, most likely from co-distillation with the solvent. Griest et al.[44] have demonstrated that the volatile PAH recovery problem may be solved by placing an adsorbent cartridge downstream of the concentration flask used in flowing-nitrogen evaporation.

Thermal methods of PAH extraction, such as sublimation, have generally received only limited attention. However, they offer certain advantages over conventional solvent-extraction methods. First, it would be expected that far less potentially interfering material would be recovered by sublimation vs. direct solvent extraction. Therefore, no further clean-up is usually required before the analysis of airborne particulate PAH. Second, concentrations of large volumes of solvent are not required.

Sublimation has been achieved both under vacuum[16,45-48] and under a nitrogen hydrocarbon carrier.[49-51] Extraction recoveries, when reported, were generally good.[79] Sollenberg[16] determined the recovery of BaP from glass fiber filters by vacuum sublimation using an isotope dilution technique. Three samples with 20 µg BaP gave 100, 102, and 103% recovery, while in two experiments with 6 ng BaP, the recovery was 95.4 and 97%, and with 0.2 ng BaP it was 91.0 and 93.8%.

For GC analysis of PAH extracted in a heated carrier gas, the sublimed PAH can be collected in a cartridge loaded with a solid adsorbent.[50] A more direct PAH extraction approach is to sublime the PAH onto the GC column.[51] In a semiquantitative analytical procedure, the PAH in strips of air particulate filters were placed in glass tubes which were inserted into a GC inlet connected with the GC column. Desorption was conducted at 350°C for 5 min. It is interesting to note that the glass capillary GC profiles of the desorbed organics were similar to those obtained from Soxhlet extracts but were achieved in a fraction of the time.

To assure accurate results, the efficiency of desorption for each PAH component from the matrix should be determined. However, relatively few studies in the literature are concerned with the determination of the efficiency of a PAH-extraction procedure. Solvent-extraction efficiencies may be tested by multiple, consecutive extractions with the same or different solvents, and by subsequent analyses of these extracts. However, in the case of highly sorptive matrices such as coal fly ash[44] or carbon black,[52] the extraction recoveries of some PAH may be very low even after numerous extractions. This may lead to serious underestimations of PAH concentrations if the recoveries are not determined otherwise.

Often, the method of spiking a particular sample matrix with known quantities of standard PAH is employed. This approach assumes that the added PAH are incorporated into the sample matrix and that they behave in the same manner as the "native" PAH. In reality, however, a uniform, homogeneous spiking of a given matrix is not easily achieved, and inhomogeneities in the spike distribution can influence apparent recoveries.[29,53] Furthermore, the assumption must be made that the PAH-extraction recoveries for similar samples are the same. This assumption may be questionable, as subtle differences among the matrices of separate samples may well lead to unexpected differences in PAH-extraction recoveries for apparently identical samples or aliquots of a given sample.

Another, more advantageous procedure employs radioactively labeled PAH tracers. The ^{14}C-labeled PAH are added to the sample, and the tracers are measured in the extract by liquid scintillation spectrometric techniques to determine extraction recoveries.[29] One disadvantage of the method is that only two tracers ([14]C- and [3]H- labeled) can be employed in a single sample because the liquid scintillation spectrometer cannot distinguish between [14]C- (or [3]H-) labels in a mixture of different PAH tracers having the same radiolabel. Multiple PAH tracers having the same radiolabel can be used if a fractionation procedure is employed to separate the tracers after extraction.[54]

B. Isolation

The extracts of the samples of airborne particles, obtained in benzene, cyclohexane, or other solvents, inevitably contain substantial quantities of compounds other than PAH which may interfere with subsequent analysis. If the analytical methods are selective enough, and the amount of co-extracted extraneous compounds low, PAH may be determined without further separation. In general, however, clean-up procedures such as solvent partitioning, column chromatography, and TLC are applied for the isolation of PAH from the extract.

1. Liquid-Liquid Partition

PAH may be enriched by partitioning the original solution with a second, immiscible solvent. If the distribution coefficients (K_D) for PAH differ from those of other, interfering materials present, they will be preferentially concentrated in one of the layers. A number of liquid-liquid extraction procedures have been reported for the isolation of PAH from air particulate extracts. Usually PAH are partitioned between a hydrocarbon solvent such as cyclohexane, isooctane, or n-alkanes, and a polar, aprotic solvent such as dimethylsulfoxide (DMSO), N,N-dimethylformamide (DMF), nitromethane, or N-methylpyrrolidone (NMP). The high solubility of these solvents toward PAH is probably due to their ability to interact with the aromatic π-system. The K_D for BaP between different solvent pairs are given in Table 4.

Extraction of PAH from cyclohexane into nitromethane has been proposed by Hoffmann and Wynder for cigarette smoke condensate.[55] This method has been incorporated into several schemes.[56-58] Usually, PAH are first separated from polar compounds by extracting the PAH cyclohexane solution with equal volumes of CH_3OH/H_2O (2:1, v/v). The PAH enriched in cyclohexane are then selectively extracted into CH_3NO_2 in a second extraction step (cyclohexane/CH_3NO_2 = 2:1, v/v) while leaving the aliphatic hydrocarbons in the cyclohexane phase. However, because of the relatively low K_D (Table 4), at least five partitions are necessary to achieve a quantitative extraction.

Grimmer and Böhnke[59] have introduced a DMF/H_2O/cyclohexane extraction method which

Table 4
K_D OF BaP FOR PAH EXTRACTION SYSTEMS

Solvent	K_D-BaP	
	Cyclohexane	Isooctane
DMSO	11	22
DMF	9.4	24
NMP	Miscible	50
Sulfolane	—	19
Propylene carbonate	8.2	—
Acetonitrile	1.3	—
Nitromethane	—	1.6
Methanol	0.82	—
Ethylene glycol	0.07	—

Adapted from Robbins, W. K., *Polynuclear Aromatic Hydrocarbons; Chemistry and Biological Effects,* Bjørseth, A. and Dennis, A. J., Eds., Battelle Press, Columbus, Ohio, 1980, 814.

has been widely used for the isolation of PAH from various environmental samples. As can be seen from Table 4, K_D are much more favorable for DMF/H$_2$O (9:1, v/v) than for nitromethane. Thus, the PAH are enriched in the DMF/H$_2$O phase, leaving aliphatic hydrocarbons in the cyclohexane layer. The PAH in the DMF/H$_2$O phase are then back-extracted into cyclohexane with a DMF/H$_2$O/cyclohexane ratio of 1:1:2 (v/v/v). The high water content in the DMF/H$_2$O phase during this second extraction step "drives" PAH into the cyclohexane phase, leaving the polar materials behind. The method has been used by Bjørseth and co-workers[37,38] in their extensive work on PAH in the occupational environment.

Another alternative solvent for the extraction of PAH from alkane solvents originally suggested by Haenni et al.,[60] is DMSO. Natusch and Tomkins[61] have made a thorough study of the use of DMSO for isolating PAH and proposed the following separation scheme for airborne particulate samples.

The PAH are dissolved in n-pentane and extracted into DMSO with the aliphatic hydrocarbons remaining in the pentane phase. An equal volume of water is then added to the DMSO phase and re-extracted with n-pentane. This allows the partitioning of PAH back into the n-pentane phase while the polar material remains in the DMSO/H$_2$O mixture. Generally, recoveries of better than 90% were obtained with this technique.[58] The method provides the advantage that the final PAH solution in n-pentane is easy to further concentrate because of its low bp. Thus, loss of volatile PAH is minimized.

Recently, Robbins[62] examined the general extraction characteristics of PAH in a number of extraction systems, DMSO, DMF, and N-methylpyrrolidone (NMP) in combination with alkane solvents. In this study, the unique properties of NMP as an extractant for PAH were demonstrated. Furthermore, the results show that PAH as a group responds systematically to changes in extraction conditions. The log of the K_D decreases linearly with (1) vol% diluent added to the extractant (2) vol% aromatic solvent present (3) the reciprocal of the absolute temperature (4) alkyl substitution on the PAH, and increases with molecular weight. This flexibility of the system allows the extraction conditions to be selected for many specific separation problems.

Even after applying the solvent-partitioning procedures, the final alkane solvent extract often contains medium polar species which may interfere with the detailed PAH analysis. Therefore, further clean-up by chromatographic methods may be needed.

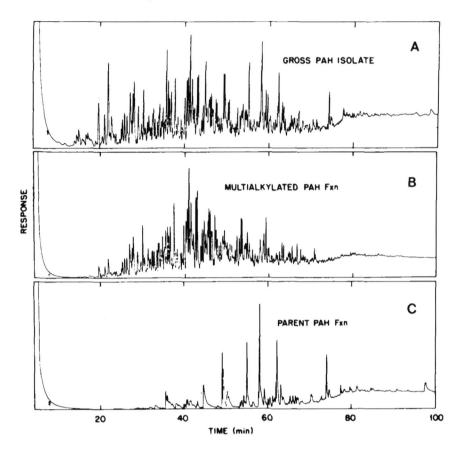

FIGURE 9.   Capillary column gas chromatogram. (A) The gross PAH isolate; (B) the multialkylated PAH fraction; (C) the parent plus simple alkylated PAH fraction. (From Griest, W. H., Tomkins, B. A., Epler, J. L., and Rao, T. K., *Polycyclic Aromatic Hydrocarbons,* Jones, P. W. and Leber, P., Eds., Ann Arbor Science Publishers, Ann Arbor, Mich., 1979, 395. With permission.)

2. Column Chromatography

The Rosen[63] type open-column liquid chromatographic (LC) separation has been widely used for fractionating complex environmental samples according to functionalities. Chromatography has been performed on alumina,[64,65] silica gel,[66,67] or Florisil®[68,69] deactivated with varying amounts of water. Elution is usually performed with an alkane solvent followed by a more polar solvent, e.g., benzene.

The main disadvantage of these classical adsorbents is the high adsorptivity which often results in loss of trace constituents and in severe peak tailing. Furthermore, reactive compounds may decompose on the active surfaces. The water content of the packing material has a strong influence on the retention volume, resulting in low reproducibility of the PAH isolation.

Lipophilic gels such as Sephadex® and Bio-Beads® have been extensively used to obtain PAH fractions for detailed analysis. Severson et al.[70] used Bio-Beads SX-12®, a neutral, porous styrene-divinylbenzene copolymer, to isolate tobacco smoke PAH. Using benzene as an eluent, multialkylated PAH can be separated from parent and monoalkylated PAH.[71] Griest et al.[72] have applied this technique to the fractionation of PAH from coal-derived crude oil into a multialkylated PAH fraction and its parent, plus a simple alkylated PAH fraction. Figure 9 demonstrates the complexity of the gross PAH isolate, resulting from the presence of numerous multialkylated PAH. The gas chromatogram of the parent plus simple alkylated PAH fraction (bottom) indicates a comparatively simple mixture.

Table 5
ELUTION VOLUMES OF PAH
ADSORBED TO 10 g SEPHADEX®-LH20

Compound	Volume (mℓ)
Aliphatic hydrocarbons	20—35
Phenanthrene/anthracene	38—50
Pyrene/fluoranthene	48—65
Benz(a)anthracene/chrysene	60—78
Benzo(b + j + k)fluoranthene	70—90
BaP	73—93
BeP	78—98
Indeno(1,2,3-cd)pyrene/perylene	81—105
Anthanthrene/benzo(ghi)perylene	89—118
Benzo(b)chrysene	84—115
Coronene	105—140

From Grimmer G. and Böhnke, H., Z. Anal. Chem., 261,
310, 1972; and Grimmer G., Ed., *Environmental Carino-
gens: Polycyclic Aromatic Hydrocarbons*, CRC Press, Inc.,
Boca Raton, Fla., 1983, 48.

The Sephadex® LH-20/isopropanol system has been applied most extensively.[59,73,74] This method is characterized by good reproducibility and high capacity. Nonpolar, nonaromatic compounds elute from the column directly after the solvent front, whereas, the PAH are retained by adsorptive forces. The PAH are then eluted in order of increasing ring number. The elution volumes of standard PAH from a Sephadex® LH-20 column with isopropanol as the mobile phase are given in Table 5.[59]

Sephadex® LH-20 has also been used as a support for the stationary phase in partition chromatography.[75,76] DMF/H_2O (85:15 v/v) is adsorbed to Sephadex® LH-20 and n-hexane, saturated with DMF/H_2O, is used as a mobile phase. As with the Bio-Bead SX-12®/benzene system, this chromatographic system allows the separation of alkylated PAH from the unsubstituted parent PAH.

3. High-Performance Liquid Chromatography (HPLC)

Recently, applications of HPLC with different stationary phases have been reported for isolating PAH from complex materials. HPLC is superior to open-column separations in terms of both speed and reproducibility. Normal-phase chromatography on silica gel[77-79] or bonded amino[80,81] silane phase has been investigated most extensively. Wise et al.[80,81] have used bonded amino columns for the isolation of PAH fractions from various samples. Separation of PAH on these columns is mainly according to the number of aromatic rings. The chromatogram of an extract from air particulates is shown in Figure 10 with the elution time of several representative PAH indicated. The PAH fractions may be collected, concentrated, and subsequently analyzed by other techniques, e.g., GC/MS. The application of high-resolution reversed-phase[82] and size-exclusion chromatography[83] has also been reported.

4. Thin-Layer Chromatography (TLC)

A TLC version of the Rosen procedure involving isolation of PAH from interfering compounds on silica gel has been used for air particulate extracts.[84] Pierce and Katz[85] applied neutral alumina TLC plates with hexane-ether (19:1 v/v), to the clean-up of airborne PAH. These methods are fast and inexpensive but have some drawbacks. Some PAH compounds are very susceptible to photooxidation when applied to silica or alumina TLC plates, and quantitative analyses, therefore, require extreme care.[86,87]

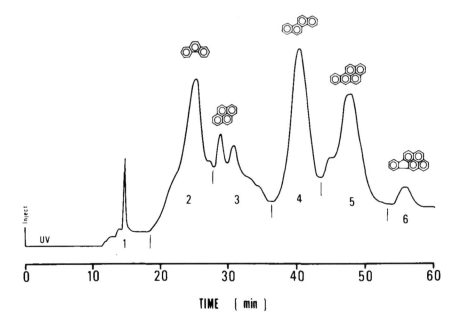

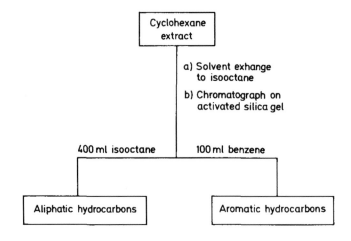

FIGURE 10.   Normal-phase liquid chromatogram of PAH extracted from urban air particulates. Column: semi-preparative μ-Bondapak NH₂ (From Wise, S. A., Bowie, S. L., Chesler, S. N., Cuthrell, W. F., May, W. E., and Rebbert, R. E., *Polynuclear Aromatic Hydrocarbons: Physical and Biological Chemistry,* Cooke, M., Dennis, A. J., and Fisher, G. L., Eds., Battelle Press, Columbus, Ohio, 1982, 919. With permission.)

FIGURE 11.   Rosen procedure as applied by Moore et al. for fractionation of PAH from air particulates.

5. Selected Analytical Schemes

In Figures 11 through 14 a number of proven schemes for the clean-up and concentration of PAH from airborne particulate matter for subsequent detailed analysis are given.

Figure 11 illustrates the simple Rosen type procedure as applied by Moore et al.[67] Bjørseth and co-workers[37,38] used a modified version of Grimmer's procedure for extensive analytical work on PAH in workplace atmospheres (Figure 12). Lee et al.[86] isolated PAH from various samples by adsorption chromatography on alumina columns followed by chromatography on Bio-Beads SX-12® as outlined in Figure 13. Recently, Grimmer et al.[73] described a simplified method for the isolation of PAH with more than three rings from air particulate

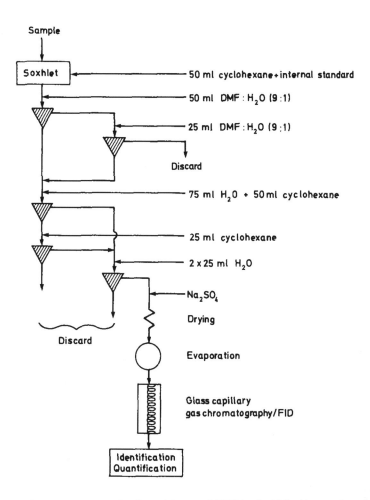

FIGURE 12. Scheme for the enrichment of PAH by liquid-liquid extraction using DMSO/H$_2$O.

matter prior to GC analysis. The scheme is given in Figure 14. Major advantages of this procedure are that it is rapid and that it can be performed in "closed systems", thereby avoiding contamination from the laboratory environment.

C. Identification and Quantification

The PAH fraction isolated from most sample types is a complex mixture of a great number of individual compounds. About 280 different PAH have been identified in cigarette smoke,[89] 146 have been identified in automobile exhaust gas,[90] and about 70 have been identified in particulate matter from work atmospheres of a coke plant and an aluminum plant.[91] Several analytical schemes have been proposed for the identification and quantification of individual PAH components. These methods usually include some type of chromatographic separation and a spectroscopic or other means of detection. Extensive reviews comparing various chromatographic procedures have been published.[92-96] The methods which show adequate separation power and sensitivity are TLC, HPLC, and GC.

1. TLC

TLC has been widely used for the separation of PAH due to its simplicity of operation, its rapidness, and the possibility of facile sample recovery for further identification of the

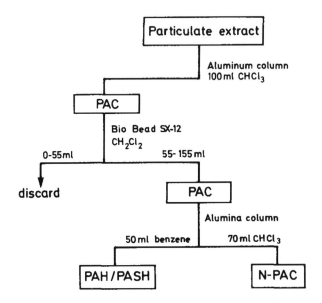

FIGURE 13.   Fractionation scheme for air particulate matter according to Lee et al.[88]

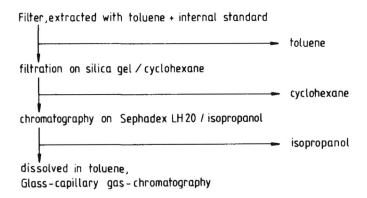

FIGURE 14.   Schematic representation of the PAH profile analysis of air-suspended matter (From Grimmer, G., Naujack, K.-W., and Schneider, O., *Fresenius Z. Anal. Chem.*, 311, 475, 1982. With permission.)

separated PAH. The principal disadvantage lies in the possibility of degradation of sensitive PAH adsorbed onto the active surface of the stationary phase.[86,87]

The use of TLC for the separation of PAH has been reviewed by several authors.[78,97-99] A great variety of different TLC systems have been described. The choice of stationary and mobile phases will depend upon the hydrocarbons to be separated and the chemical characteristics of the background material from which they are to be isolated. The three most commonly used TLC systems have been compared by Sawicki et al.[97] These are (1) alumina with pentane-diethylether (19:1), (2) cellulose acetate with ethanol-toluene-water (17:4:4), and (3) cellulose with dimethylformamide-water (1:1). Table 6 shows the retention values of various PAH relative to BaP. Although alumina gave best separation of PAH from other compound types, the PAH were poorly separated from each other. The cellulose system gave the widest range of retention values, and the separation of difficult pairs could often be improved with a slight change of water content in the mobile phase. The cellulose acetate

Table 6
R_B VALUES OF PAH (R_B OF BaP = 1.00)

Compound	Systems[a]		
	1	2	3
Phenanthrene	1.13	3.74	1.99
Anthracene	1.14	3.33	1.99
Fluoranthene	1.09	2.92	1.89
Chrysene	1.10	—	1.75
Pyrene	1.25	3.16	1.72
Triphenylene	1.07	—	1.49
Benz(a)anthracene	1.03	2.70	1.47
11H-Benzo(b)fluorene	1.08	3.54	1.33
BeP	1.04	2.94	1.16
Perylene	0.19	2.86	1.14
Benzo(k)fluoranthene	0.98	2.40	1.03
BaP	1.00	1.00	1.00
Anthanthrene	0.71	2.17	0.70
Benzo(ghi)perylene	0.89	3.04	0.69
Dibenz(a,h)anthracene	0.74	2.92	0.66
Naphthol(1,2,3,4-def)chrysene	0.78	1.85	0.48
Benzo(rst)pentaphene	0.68	2.41	0.45
Coronene	0.46	2.87	0.37
Benzo(a)coronene	0.10	2.48	0.15
Dibenzo(h,rst)pentaphene	0.12	2.35	0.14

[a] See text.

Adapted from Sawicki, E., Stanley, T. W., Elbert, W. C., and Pfaff, J. D., *Anal. Chem.*, 36, 497, 1964.

system gave the best separation of the five-ring PAH fraction, including BaP, BeP, benzofluoranthenes and perylene.

Poor resolution and sensitivity are not inherent to TLC. They merely reflect the fact that TLC, as normally practiced, is not a high-performance technique. In recent years efforts have been made to improve the efficiency and sensitivity of TLC.[100]

Two-dimensional development in TLC has been used and is particularly suitable for mixtures of many components such as are normally found in PAH fractions. Matsushita[48] described a two-dimensional dual-band TLC system. The TLC plate consists of an aluminum oxide layer (4 × 20 cm) and a 26% acetylated cellulose layer (18 × 20 cm). The PAH extract is applied onto the aluminum oxide layer and developed with n-hexane-ether (19:1, v/v) to the 15 cm mark on the aluminum oxide layer. PAH are then separated on the acetylated cellulose layer using methanol-ether-water (4:4:1, v/v/v). PAH may be quantified by in situ scanning with a fluorescence scanning densitometer. Figure 15 shows the thin-layer chromatogram for a sample of airborne particulate matter.

Recently, Tomingas and Grover[101] reported the use of reversed-phase high-performance TLC for the rapid analysis of a number of PAH frequently found in ambient air. The method had a good reproducibility and high sensitivity with a minimum detectable limit of 0.2 ng for BaP.

2. HPLC

Since its inception in the early 1970s, HPLC using microparticulate packings has been widely used for the separation of PAH. An extensive review has been given by Wise.[102] HPLC has several advantages for analyzing PAH. First, HPLC offers a variety of stationary

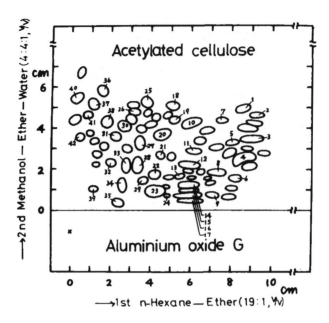

FIGURE 15. Two-dimensional dual-band thin-layer chromatogram of vacuum sublimation extract of airborne particulates. Identified PAHs and their spot number on the chromatogram are as follows: 5,12-dihydrotetracene (spot 2), 3-methylpyrene and 7H-benzo(c)fluorene (3), pyrene and fluoranthrene (4), benzo(mno)fluoranthene (5), benz(a)anthracene (8), chrysene (9), 13H-naphtho(2,3-b)fluorene (10), BeP (11), perylene (12), benzo(k)fluoranthene (14), benzo(j)fluoranthene (15), benzo(b)fluoranthene (16), BaP (17), benzo(ghi)perylene (20), 11H-naphtho(2,1-a)fluorene (22), indeno(1,2,3-cd)pyrene (23), anthanthrene (24), coronene (26), picene and tetracene (27), benzo(b)chrysene (29), tribenzo(a,c,j)tetracene (30), tribenzo(a,e,i) pyrene (33), dibenzo(a,i)pyrene (34), dibenzo(a,h)pyrene and benzo(g)chrysene (35), 1,2,3,5,6,7-hexahydrotriangulene and naphtho(2,3-k)fluoranthene (38). (From Matsushita, H., *Polycyclic Hydrocarbons and Cancer*, Vol. 1, Gelboin, H,. V. and Ts'o, P. O. P., Eds., Academic Press, New York, 1978, 71. With permission.)

phases capable of providing unique selectivity for the separation of PAH isomers. Second, UV absorption and fluorescence spectroscopy provide highly sensitive and selective detection for PAH in HPLC. Finally, HPLC provides a useful fractionation technique for the isolation of PAH for subsequent analysis by other techniques.

Reverse-phase HPLC on chemically bonded nonpolar phases such as C_{18} is by far the most popular LC method for the separation of PAH. The reverse-phase mode provides unique selectivity for the separation of PAH isomers that are often difficult to separate by other modes of LC. In addition, the compatibility of reverse-phase HPLC with gradient elution techniques, and the rapid equilibration of the columns to new mobile-phase compositions make reverse-phase HPLC a convenient separation technique for PAH. The separation of a number of PAH on two different 10 μm C_{18} columns is shown in Figure 16.

Several recent reports have compared retention data of PAH on various columns.[81,103-107] These studies indicate that C_{18} columns from various manufacturers do not only provide different separation efficiencies, but also quite different selectivities and retention characteristics for PAH. Besides other factors such as surface area and carbon content, the nature of the organic layer is of importance for the selective retention of PAH. The chromatograms shown in Figure 16 were obtained on columns with a monomeric C_{18} layer and a polymeric

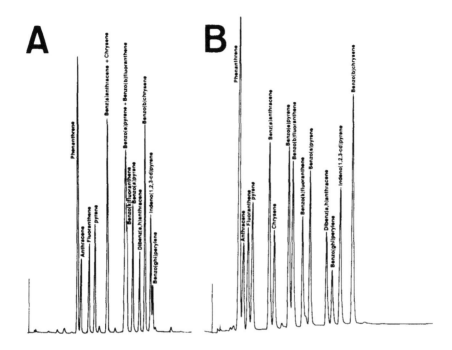

FIGURE 16. Comparison of reverse-phase HPLC separation of PAH on two different C_{18} columns. (A) Zorbax ODS, monomeric; (B) Vydac 201 TP, polymeric. (From Bartle, K. D., Lee, M. L., and Wise, S. A., *Chem. Soc. Rev.*, 10, 113, 1981. With permission.)

C_{18} layer, respectively. Generally, better separation of critical isomers is obtained on polymeric C_{18} phases. Parameters such as composition of mobile phase[108] and column temperature[109] can be used to achieve changes in the elution order of PAH in reverse-phase HPLC.

Polar chemically bonded stationary phases used in conjunction with nonpolar mobile phases (normal-phase HPLC), have been employed for the separation of PAH. Several polar phases are available containing functional groups such as amine (NH_2), diamine ($R(NH_2)_2$), nitrile (CN), diol ($R(OH)_2$), ether (ROR), and nitrophenyl (NO_2) bonded to silica particles. In the normal-phase mode, the PAH separations achieved are similar to those obtained on classical adsorbents such as alumina and silica. Lankmayr and Müller[110] compared NO_2-, NH_2-, and C_{18}-columns for the separation of 17 PAH, and found the best separation on the NO_2 phase. Recently, Chemilowiec and George[111] investigated the performance of NH_2, CN, $R(OH)_2$, ROR, and $R(NH_2)_2$ columns for normal-phase separation of ten PAH. These authors suggested that the diamine column ($R(NH_2)_2$) was superior to the other polar-bonded phases for PAH separations.

Recently, Sonnefeld et al.[112] described an on-line multidimensional HPLC method for PAH in complex samples. The desired fractions obtained from a semi-preparative amino column are concentrated on a bonded diamine column (concentrator column). The normal-phase solvent is evaporated using an inert gas purge. The concentrator column is then connected to the reverse-phase analytical system, and the fraction is transferred onto the analytical C_{18} column using gradient elution focusing. This method permits rapid qualitative and quantitative determinations of PAH in complex mixtures.

HPLC is particularly suitable for the analysis of high molecular weight PAH (>300 daltons). With a reverse-phase C_{18} column, Peaden et al.[113] identified numerous PAH up to a molecular weight of 448 (11 rings) with a gradient including dichloromethane as final eluent. However, as the number of isomers increases drastically with the increasing number of aromatic rings, resolution of conventional HPLC becomes insufficient. Recently developed

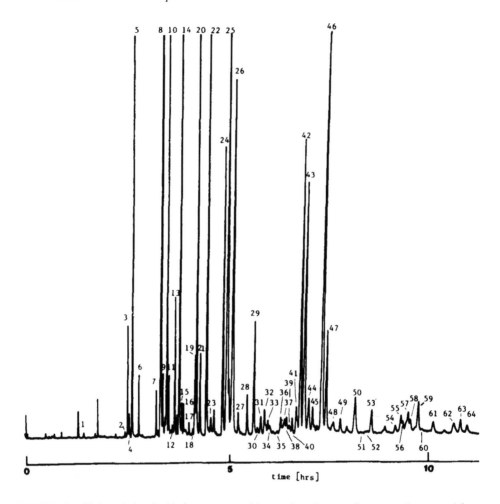

FIGURE 17. High-resolution liquid chromatogram of large polycyclic aromatic compounds extracted from carbon black. (From Hirose, A., Wiesler, D., and Novotny, M., *Chromatographia*, 18, 239, 1984. With permission.)

microbore columns[114] (I.D. 1 to 2 mm) and HPLC capillary columns[115] (I.D. 5 to 200 μm) may successfully improve the resolving power. Hirose and co-workers[116] have demonstrated the high degree of resolution obtained in the separation of large PAH using a 180 cm × 200 μm I.D. fused silica capillary column packed with 3 μm C_{18}-bonded silica. Figure 17 shows the separation of large PAH (six- to nine-ring structures) from an extract of carbon black.

While HPLC systems for microbore columns are commercially available, capillary HPLC is still under development. Apart from current instrumental limitations of small-volume technology[117] in capillary HPLC, long separation times are a major difficulty. These problems may be overcome by using a supercritical fluid as the mobile phase (SFC). Above the critical temperature, the "solvent" is compressed into a fluid with a density similar to that of a liquid, but with a low viscosity similar to that of a gas. Solubility of solutes in the supercritical mobile phase is generally good. Therefore, this technique offers a potential for the rapid, efficient separation of nonvolatile PAH.

SCF chromatography may be used with capillary columns with immobilized stationary-phase film,[118] or with conventional packed columns.[119] Peaden et al.[120] demonstrated that efficient separation of coal tar components is obtained using a SCF capillary system with pressure programmed supercritical n-pentane.

A major advantage of HPLC for the determination of PAH is the availability of sensitive and selective detectors. The UV detector operated at 254 nm is universal for PAH, as all PAH exhibit some absorption at this wavelength. However, it is usually possible to improve the sensitivity by monitoring at other wavelengths. Several authors[121,122] have described the use of variable-wavelength UV detection to achieve some degree of selectivity. Absorbance ratios at several wavelengths have been employed to identify chromatographic peaks.[121-123]

Fluorescence detection of PAH provides much greater sensitivity and selectivity than UV detection. The specificity of fluorescence allows for the selective determination of PAH in the presence of nonfluorescing interferences, often avoiding the need for sample clean-up. Another advantage of fluorescence detection is the selectivity achieved for individual PAH and the possible identification of specific PAH in complex mixtures. This is illustrated in Figure 18 which compares UV detection with fluorescence selective for perylene (Peak A, $\lambda_{ex} = 405$, $\lambda_{em} = 440$ nm), and with more universal fluorescence conditions ($\lambda_{ex} = 300$, $\lambda_{em} = 400$ nm). May and Wise[124] optimized the selectivity for individual PAH by changing the fluorescence excitation and emission wavelengths during the reverse-phase chromatographic separation. The selectivity achieved by using this approach is demonstrated in Figure 19. The spectrofluorimetric detector was programed to change wavelength conditions nine times during the chromatographic run, resulting in excellent resolution of ten major PAH.

The fluorescence spectral characteristics for a great number of PAH have been summarized.[102,125] Ogan et al.[126] compared the use of a cut-off filter and a monochromator on the emission side in the fluorescence detection of PAH in environmental samples. The filter fluorimeter provided an improvement of three to five times in sensitivity over the monochromator; however, these gains were often offset by spectral interferences from other compounds due to lower selectivity of the filter instrument. Thus, the authors recommend the use of monochromators to reduce spectral interferences from overlapping peaks. When using fluorescence detection in HPLC, the mobile phase should be deoxygenated to avoid fluorescence quenching of PAH.[127]

The potential of the combination of liquid chromatography-mass spectrometry (LC-MS) for the separation and identification of organic compounds has generated considerable research in the past few years. The inherent problem of introducing a liquid stream into the high-vacuum system of the MS has led to several approaches which have been reviewed in the literature.[128,129] Applications of LC-MS for the identification of low-volatile compounds, including PAH, have been reported.[129,130]

3. GC

GC has been used in the separation of PAH since the late 1950s and early 1960s,[131,132] and most of the work was done on packed columns. The liquid stationary phases most commonly employed are listed in Table 7.

The carborane-silicone polymers appear highly suitable for packed-GC analysis of PAH.[135] Low column bleeding and good separation efficiency have been demonstrated on a variety of environmental samples.[135,140,141] Nematic liquid crystals have also been used as a stationary phase in the GLC of PAH.[142,143] These phases have excellent resolution of some PAH isomers such as chrysene and triphenylene, which are hard to separate on other stationary phases. However, column bleeding at higher temperatures remains a problem with these phases.

Recently, the method of capillary gas chromatography (CGC)[144-148] has been developed as an extremely useful tool in characterizing multicomponent mixtures. The CGC exhibits excellent reproducibility, high sensitivity, and good resolution among individual isomeric PAH when compared to regular packed-column GC.[149] This method appears to be the most satisfactory one for PAH measurement.[59,93-95,149-151]

Figure 20 shows a chromatogram of coal tar PAH obtained with a persilanized glass capillary column coated with OV-73. The performance of this column as measured by the

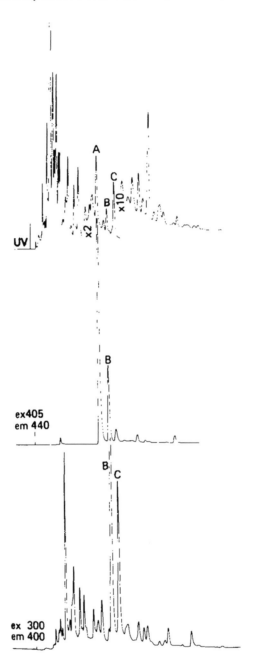

FIGURE 18.   HPLC separation of PAH from coal tar extract. (1) UV-detection at 254; (2) fluorescence at λex = 405 and λem = 440 nm; (3) fluorescence at λex = 300 and λem = 400 nm. (From Wise, S. A., *Handbook of Polycyclic Aromatic Hydrocarbons*, Bjørseth, A., Ed., Marcel Dekker, New York, 1983, 183. With permission.)

resolution of isomeric pairs phenanthrene (16)/anthracene (17), benz(a)anthracene (39)/chrysene (40), and BeP (45)/BaP (46) represents the best resolution currently available. Glass, and most recently, fused-silica[152] columns are now used universally. To avoid peak tailing and adsorption of trace compounds, several methods of surface treatments have been employed.[153]

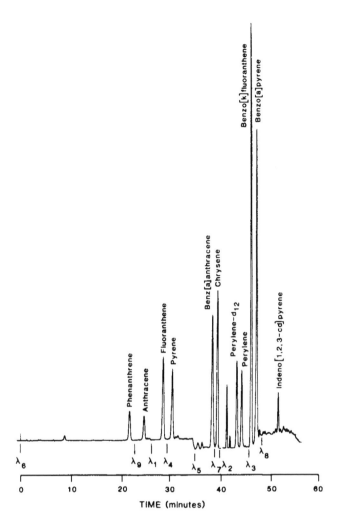

FIGURE 19.   Reverse-phase HPLC analysis of PAH from air particulate extract using fluorescence wavelength programing (From May, W. E. and Wise, S. A., *Anal. Chem.*, 56, 225, 1984. With permission.)

Table 7
LIQUID STATIONARY PHASES FOR GC
SEPARATION OF PAH

Chemical type	Commercial names	Ref.
Methylsilicone fluid	OV-101, SP-2100	133
Methylsilicone gum	SE-30, OV-1	134, 135
Methylphenylsilicone fluid	SE-52, (5% phenyl)	136
	OV-17 (20% phenyl)	137
	OV-25 (75% phenyl)	69
Carborane/silicone polymer	Dexsil 300	134, 135, 138
	Dexsil 400	135
Polyphenylether sulfone	Poly S 179	139

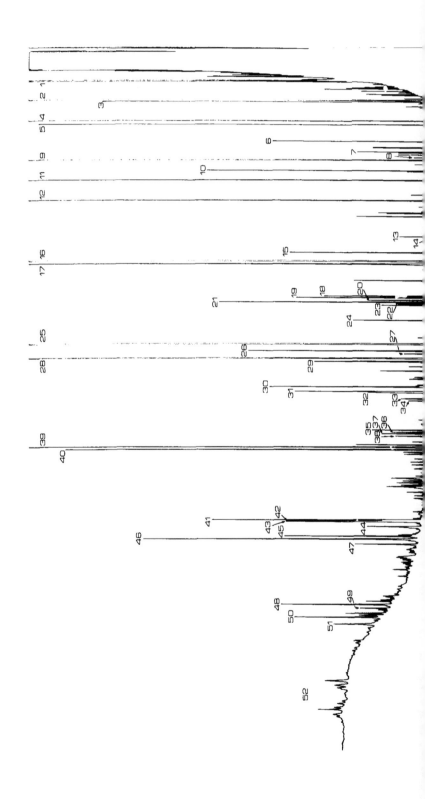

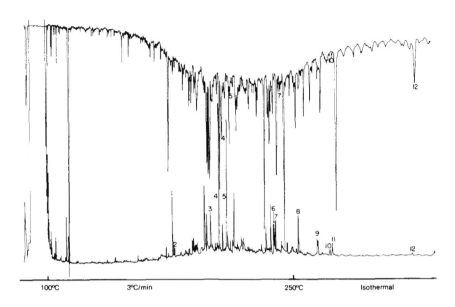

FIGURE 21. Glass capillary gas chromatogram of PAH in air particulates with simultaneous FID (lower trace) and ECD. (From Bjørseth A. and Eklund, F., *J. High Resol. Chromatogr. Chromatogr. Commun.*, 1, 22, 1979. With permission.)

A variety of stationary phases have been used in capillary column separation of PAH;[151] however, because of their high efficiency and high-temperature stability, silicone gum phases such as SE-52 and SE-54 have gained widest acceptance.

For most PAH analytical work, columns need be no longer than 10 to 25 m with I.D. of 0.2 to 0.3 mm and film thicknesses near 0.3 μm. This shortens analysis time and allows for the elution of larger PAH.[154] Hydrogen has been established as the ideal carrier gas in CGC, as it results in higher separation efficiency, shorter elution times, and better sensitivity.[155]

The sample-introduction technique is critical when mixtures of compounds differing widely in volatility, such as PAH, are chromatographed on capillary columns. A comparison of various injection techniques with regard to discrimination and decomposition effects has been published.[156] In general, the nonsplitting techniques are much more desirable than those involving splitting because of the small amount of PAH usually present in environmental samples. The development of the on-column injector[157] has essentially eliminated discrimination caused by differences in volatility, polarity, or concentration. However, as the sample is introduced directly into the capillary column, it must be relatively free of nonvolatile material to avoid column deterioration.

The most widely used GC detector for PAH is the FID. It has several advantages: (1) the high linearity range makes it ideal for quantitative work based on internal standards, (2) there is usually no need for correcting for different response factors in practical routine work, and (3) it is reliable and easy to maintain. Its general response character, however, necessitates extensive clean-up procedures prior to GC to eliminate possible interfering compounds.

Other detectors, such as the electron capture detector (ECD), have also proved useful for specific purposes.[94,158] Bjørseth and Eklund[158] investigated the possibility of simultaneous detection with an ECD and an FID for analyzing PAH. The ECD/FID ratios were determined for 46 compounds and the application to analysis of airborne PAH demonstrated. An example of the work is given in Figure 21. The results suggest that the method may be used for obtaining additional evidence in identifying PAH.

The most powerful approach available today for the analysis of complex PAH mixtures is the use of a mass spectrometer as detector in capillary gas chromatography (GC/MS).[91,159] The high resolving capacity of CGC, together with MS molecular structure information, often leads to complete identification of PAH components in complex mixtures. Figure 22 shows the total ion current chromatogram of the PAH fraction in particulates from the working atmosphere of a coke plant.[91] By comparing retention times with standards combined with mass spectral information, more than 100 compounds could be detected and identified.

Electron impact ionization (EI) often leads to identical fragmentation of isomers, and thus to identical mass spectra. Figure 23 gives the EI mass spectra for the two isomeric four-ring PAH, fluoranthene and pyrene. The two mass spectra are indistinguishable within the experimental variations normally encountered when analyzing real samples. New ionization methods that distinguish among isomeric PAH promise increased diagnostic power in the future.[160]

IV. QUALITY CONTROL OF THE ANALYTICAL METHOD

In chemical analysis, determination of accuracy and precision is essential. Accuracy is a measure of how close a determined value comes to the true value; precision determines how well this value is reproduced. Furthermore, when determining PAH in trace levels, it is important to consider whether or not the correct analyte was measured. In real-world samples, qualitative results are often as difficult to obtain as quantitative results.[161]

A. Use of Internal Standards

The measurement of, and the correction for, all systematic errors in the analytical procedure is a tedious and often impossible process. One way to circumvent much of this problem is to use an internal standard, which should be added to the sample as early as possible in the analytical scheme. The internal standard is assumed to be as susceptible to the same systematic errors during the analyses as the compounds being determined. By measuring the relative signals for the internal standard and for the unknown PAH (or any other compound), the concentration of the unknown can be calculated. This supposition is valid only if the internal standard exhibits all the chemical and physical properties of the organic compound being determined, and only if it is present in approximately the same concentration as the compound of interest.

The use of internal standards in GC analysis of PAH has gained wide acceptance. As the PAH fraction covers a bp range of more than 300°C, there is a need for more than one internal standard (to correct for evaporation losses and peak broadening during a chromatographic run). Bjørseth et al.[37,38,149] routinely use the two compounds 3,6-dimethylphenanthrene and 2,2′-binaphthyl, each of which covers one part of the temperature-programed GC run. Other PAH such as benzo(a)chrysene or indeno(1,2,3-cd)fluoranthene have been used as internal standards.[73] Grimmer et al.[73] showed that the FID response factor of 15 PAH, ranging from fluoranthene to coronene, varied from 0.993 to 1.055 relative to the internal standard indeno(1,2,3-cd)fluoranthene. Therefore, the relative peak areas gave a good representation of the quantitative composition of the PAH mixture isolated from the sample.

Using naphthalene, phenanthrene, chrysene, and picene as internal standards, Vassilaros et al.[162] have proposed a linear retention index system for temperature-programed CGC of PAH. Usually, these compounds are indigenous to the sample and not added. They serve to calculate retention indexes in order to facilitate the identification of PAH compounds in complex mixtures. A similar system has been introduced by Wise[102] and co-workers for reverse-phase HPLC of complex PAH mixture based on the elution volumes of benzene, naphthalene, phenanthrene, benz(a)anthracene, and benzo(b)chrysene.

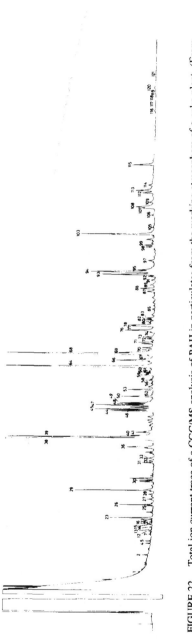

FIGURE 22. Total ion current trace of a CGC/MS analysis of PAH in particulates from the working atmosphere of a coke plant. (From Bjørseth, A. and Eklund, G., *Anal. Chim. Acta*, 105, 199, 1979. With permission.)

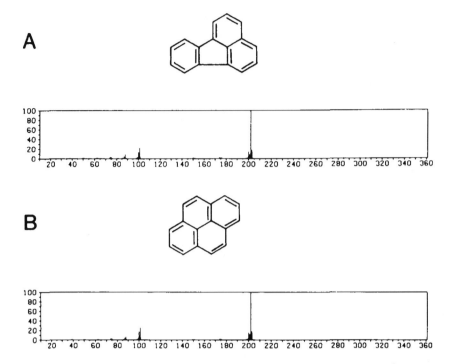

FIGURE 23.   EI mass spectra of (A) fluoranthene and (B) pyrene. (From Lee, M. L., Novotny, M. V., and Bartle, K. D., *Analytical Chemistry of Polycyclic Aromatic Compounds*, Academic Press, New York, 1981. With permission.)

B. Intra- and Interlaboratory Control of PAH Analyses

The confidence one can place on a measurement by a given laboratory depends both on how close the laboratory comes, on the average, to the true value, and how much the individual measurements vary. For complex analyses such as the determination of PAH, there may exist a number of sources for the variations, such as:

1.   The method used for extraction, clean-up, and analysis
2.   The experience and skill employed in using the methods
3.   The amount of PAH in the sample

It has been pointed out that analytical laboratories, generally, are more precise than accurate. Consequently, the variation of measurements within a given laboratory is small compared to the variation observed in interlaboratory ("round robin") studies.[163] This is demonstrated in Tables 8 and 9. The variation in the gas chromatographic PAH analysis of a vehicle exhaust condensate performed ten times by the same laboratory is given in Table 8.[164] The relative standard deviation (RSD) for the various PAH varies between 2.8 and 5.8%. Table 9 shows results from an interlaboratory comparison of the analysis of PAH in exhaust condensate using independent methods.[164] Here, the variation of the analytical results is much larger, with the RSD ranging from 7.2 to 48.7%.

Similar results were obtained from a round robin test performed in the Nordic countries on electrofiltered particles from an aluminum plant.[165] While the repeatability of the PAH analyses was generally good, with a mean RSD between 4.1 and 21.5%, the interlaboratory variations ranged from 22.5% for BaP to 45.2% for BeP.

These examples reveal the relatively large systematic biases among different laboratories, and document the current state-of-the-art for routine PAH measurements. One method for

Table 8
TENFOLD PAH ANALYSIS (GC) OF EXHAUST CONDENSATE (TO DETERMINE VARIATION COEFFICIENTS)

	Analysis no. (values in µg)										Var. Coeff. %
	1	2	3	4	5	6	7	8	9	10	
Fluorene	263.0	247.0	271.0	248.0	238.0	258.0	224.8	250.0	241.0	259.0	5.4
Pyrene	468.0	442.0	486.0	442.0	430.0	463.0	428.0	455.0	428.0	470.0	4.5
Benzo(ghi)fluoranthene	93.0	98.0	94.0	92.6	95.8	96.0	93.0	95.6	94.5	97.0	3.1
Chrysene-cyclopenta(cd)pyrene	307.0	289.0	314.0	320.0	285.0	316.0	306.0	320.0	305.0	322.0	4.2
Benzo(b+j+k)fluoranthene	39.8	38.8	38.9	37.7	37.0	41.2	40.9	42.0	40.8	41.3	3.8
BeP	35.2	33.7	34.3	38.0	35.5	36.8	37.0	37.6	38.0	38.4	4.7
BaP	39.3	38.1	38.3	41.0	40.4	40.2	39.6	41.1	40.2	41.5	2.8
Perylene	6.2	6.1	5.8	6.1	6.0	6.3	6.1	6.5	6.2	6.3	3.2
Indeno(1,2,3-cd)pyrene	33.9	33.4	33.1	32.6	33.1	35.8	33.6	34.9	33.6	35.0	2.9
Benzo(ghi)perylene	137.0	132.0	133.0	140.0	144.2	150.0	143.0	150.0	145.0	150.0	4.8
Coronene	105.0	105.5	105.0	111.0	111.0	121.0	105.8	122.0	111.0	116.0	5.8

From Janssen, T., *Luftverunreinigungen durch polycyclische aromatische Kohlenwasserstoffe — Erfassung und Bewertung*, VDI-Berichte 358, VDI-Verlag, Düsseldorf, 1980, 69. With permission.

Table 9
ROUND ROBIN TEST: EXHAUST
CONDENSATE

Compounds	N	X	S	V (%)
Fluoranthene	27	432	31.1	7.2
Pyrene	28	725	57.8	7.9
Chrysene	35	261	127.5	48.7
Benzofluoranthene	35	52.1	10.9	20.9
BeP	33	36.8	3.7	10.2
BaP	31	44.0	6.6	15.0
Perylene	28	8.8	1.8	22.5
Benzo(ghi)perylene	34	156.2	19.5	12.5
Coronene	37	101.2	24.3	24.0

Note: N = number of values, X = average, S = standard
deviation, V (%) = variation coefficients.

From Janssen, T., *Luftverunreinigungen durch polycyclische aromatische Kohlenwasserstoffe — Erfassung und Bewertung,* VDI-
Berichte 358, VDI-Verlag, Düsseldorf, 1980, 69. With permission.

enhancing accuracy is the frequent analyses of suitable standard reference materials under careful control of all experimental parameters. Similarly, contamination problems should be controlled by frequent analyses of blanks.

V. SUMMARY AND CONCLUSION

There exist numerous methods which may be used to determine PAH in workplace atmospheres. Traditionally, the benzene-soluble fraction method has been widely used to evaluate exposure to coal tar pitch volatiles. However, this method has the drawback that it is nonspecific and subject to a number of interferences. Therefore, several chromatographic techniques have been employed to separate PAH from the benzene-soluble fraction and to determine PAH as a group.

Several methods have been developed for the specific determination of BaP as a proxy compound representing the large and complex group of PAH compounds present in work atmospheres.

Most methods for detailed PAH analysis are basically three-step procedures involving extraction and isolation of PAH, and separation and quantitation of individual PAH compounds. Care should be taken to quantitatively extract the PAH from the sample, and to avoid losses of PAH during clean-up. For the detailed characterization of multicomponent PAH mixtures, analytical techniques with high separation efficiency and high sensitivity are required. In this respect, capillary GC is regarded as the most powerful tool. The combination with a mass spectrometer gives in addition molecular structure information which often leads to complete characterization of complex samples.

Even though HPLC does not provide the high separation efficiency of CGC, it has recently been developed into an excellent method for measuring individual PAH in complex mixtures. HPLC provides a great variability in selective separations of PAH isomers, and allows highly sensitive and selective detection of individual PAH by fluorescence spectrometry. Finally, normal-phase HPLC provides a useful fractionation and clean-up technique for the isolation of PAH from complex mixtures.

TLC has been widely used because of its simplicity and rapidity of operation, and because of the possibility of easy sample recovery. When efforts are made to improve the efficiency of TLC, the resulting improvements in resolution and sensitivity make high-performance TLC comparable to HPLC.

Interlaboratory comparisons show that, generally, good precision can be obtained in PAH analyses. Accuracy, however, may vary considerably. Round robin tests using independent methods show that the RSD for the determination of some PAH may be up to 40 to 50%. Therefore, frequent quality controls of PAH analyses by analyzing reference samples and determining blank values are recommended.

REFERENCES

1. **Hawthorne, A. R., Thorngate, J. H., Gammage, R. B., and Vo-Dinh, T.,** Development of a prototype instrument for field monitoring of PAH vapors, in *Polynuclear Aromatic Hydrocarbons,* Jones, P. W. and Leber, P., Eds., Ann Arbor Science Publ., Ann Arbor, Mich., 1979, 299.
2. **Hawthorne, A. R. and Thorngate, J. H.,** Application of second-derivative UV-absorption spectroscopy to polynuclear aromatic compound analysis, *Appl. Spectrosc.,* 33, 301, 1979.
3. **Vo-Dinh, T., Gammage, R. B., and Martinez, P. R.,** Identification and quantification of polynuclear aromatic compounds in Synthoil by room-temperature phosphorimetry, *Anal. Chim. Acta,* 118, 313, 1980.
4. **Vo-Dinh, T., Gammage, R. B., and Martinez, P. R.,** Analysis of a workplace air particulate sample by synchronous luminescence and room-temperature phosphorescence, *Anal. Chem.,* 53, 253, 1981.
5. **Vo-Dinh, T.,** Synchronous excitation spectroscopy, in *Modern Fluorescence Spectroscopy,* Vol. 4, Wehry, E. L., Ed., Plenum Press, New York, 1981, 167.
6. **Lane, D. A., Sakuma, T., and Quan, E. S. K.,** Real-time analysis of gas phase polycyclic aromatic hydrocarbons using a mobile atmospheric pressure chemical ionization mass spectrometer system, in *Polynuclear Aromatic Hydrocarbons: Chemistry and Biological Effects,* Bjørseth, A. and Dennis, A. J., Eds., Battelle Press, Columbus, Ohio, 1980, 199.
7. **Sakuma, T., Davidson, W. R., Lane, D. A., Thomson, B. A., Fulford, J. E., and Quan, E. S. K.,** Compounds in hot gas streams by APCI/MS and APCI/MS/MS, in *Chemical Analysis and Biological Fate: Polynuclear Aromatic Hydrocarbons,* Cooke, M. and Dennis, A. J., Eds., Battelle Press, Columbus, Ohio, 1981, 179.
8. **Fulford, J. E., Sakuma, T. E., and Lane, D. A.,** Real-time analysis of exhaust gases using triple quadrupole mass spectrometry, in *Polynuclear Aromatic Hydrocarbons: Physical and Biological Chemistry,* Cooke, M., Dennis, A. J., and Fisher, G. L., Eds., Battelle Press, Columbus, Ohio, 1982, 297.
9. **Seim, H. J., Hannemann, W. W., Barsotti, L. R., and Walker, T. J.,** Determination of pitch volatiles in airborne particulates, *Am. Ind. Hyg. Assoc. J.,* 35, 718, 1974.
10. **MacEachen, W. L., Boden, H., and Larivière, C.,** Composition of benzene soluble matter collected at aluminum smelter workplaces, *Light Met. (New York),* 2, 509, 1978.
11. **Bonney, T. B.,** Measurement of polycyclic aromatic hydrocarbons in the aluminum industry, presented at the 2nd IPAI Health Protection Semin., Montreal, September 22 to 24, 1981.
12. NIOSH Criteria for a Recommended Standard: Occupational Exposure to Coal Tar Products, National Institute for Occupational Safety and Health, U.S. Department of Health, Education and Welfare, Cincinnati, 1977.
13. **Boden, H. and Roussel, R.,** A method for determining polycyclic aromatic hydrocarbons for coal tar pitch volatiles, *Light Met. (Warrendale, Pa.),* 1049, 1983.
14. **Balya, D. R. and Danchik, R. S.,** Chromatographic techniques for characterization of coal tar pitch volatiles, *Am. Ind. Hyg. Assoc. J.,* 45, 260, 1984.
15. **Fjeldstad, P. E. and Halgard, K.,** A Simple Thin-Layer Chromatographic Method for Routine Analysis of Polycyclic Aromatic Hydrocarbons (in Norwegian), HD 696, Institute of Occupational Health, Oslo, 1976.
16. **Sollenberg, J.,** A method for determining benzo(a)pyrene in air samples collected on glass fiber filters in occupational areas, *Scand. J. Work Environ. Health,* 3, 185, 1976.
17. **Blome, H.,** Messungen polycyclischer aromatischer Kohlenwasserstoffe an Arbeitsplätzen — Beurteilung der Ergebnisse, *Staub-Reinhalt. Luft,* 41, 225, 1981.
18. **Boden, H.,** The determination of benzo(a)pyrene in coal tar pitch volatiles using HPLC with selective UV detection, *J. Chromatogr. Sci.,* 14, 391, 1976.
19. **Tomkins, B. A., Reagan, R. R., Caton, J. E., and Griest, W. H.,** Liquid chromatographic determination of benzo(a)pyrene in natural, synthetic and refined crudes, *Anal. Chem.,* 53, 1213, 1981.

20. **Colmsjö, A. and Stenberg, U.,** Identification of polynuclear aromatic hydrocarbons by Shpol'skii low temperature fluorescence, *Anal. Chem.,* 51, 145, 1979.

21. **Yang, Y., D'Silva, A. P., Fassel, V. A., and Iles, M.,** Direct determination of polynuclear aromatic hydrocarbons in coal liquids and shale oil by laser excited Shpol'skii spectrometry, *Anal. Chem.,* 52, 1350, 1980.

22. **Yang, Y., D'Silva, A. P., and Fassel, V. A.,** Deuterated analogues as internal reference compounds for the direct determination of benzo(a)pyrene and perylene in liquid fuels by laser-excited Shpol'skii spectrometry, *Anal. Chem.,* 53, 2107, 1981.

23. **Smith, E. M. and Levins, P. L.,** Sensitized fluorescence detection of PAH, in *Polynuclear Aromatic Hydrocarbons: Chemistry and Biological Effects,* Bjørseth, A. and Dennis, A. J., Eds., Battelle Press, Columbus, Ohio, 1980, 973.

24. International Agency for Research on Cancer, *IARC Monographs on the Evaluation of the Carcinogenic Risk of Chemicals to Humans,* Polynuclear aromatic compounds. I. Chemical, environmental and experimental data, Vol. 32, IARC, Lyon, France, 1983.

25. **Dipple, A.,** Polynuclear aromatic carcinogens, in *Chemical Carcinogens,* Searle, C. E., Ed., American Chemical Society, Washington, D.C., 1976, 245.

26. **Hecht, S. S., Amin, S., Rivenson, A., and Hoffmann, D.,** Tumor initiating activity of 5,11-dimethylchrysene and the structural requirements favoring carcinogenicity of methylated polynuclear aromatic hydrocarbons, *Cancer Lett.,* 8, 65, 1979.

27. **Gammage, R. B. and Bjørseth, A.,** Proxy methods and compounds for workplace and monitoring of polynuclear aromatic hydrocarbons, in *Polynuclear Aromatic Hydrocarbons: Chemistry and Biological Effects,* Bjørseth, A. and Dennis, A. J., Eds., Battelle Press, Columbus, Ohio, 1980, 565.

28. **Schaad, R. E.,** Chromatography of (carcinogenic) polycyclic aromatic hydrocarbons, *Chromatogr. Rev.,* 13, 61, 1970.

29. **Griest, W. H. and Caton, J. E.,** Extraction of polycyclic aromatic hydrocarbons for quantitative analysis, in *Handbook of Polycyclic Aromatic Hydrocarbons,* Bjørseth, A., Ed., Marcel Dekker, New York, 1983, 95.

30. **Chatot, G., Castegnaro, M., Roche, J. L., Fontanges, R., and Obaton, P.,** Comparative study of ultrasonic and soxhlet extraction of atmospheric polycyclic hydrocarbons, *Anal. Chim. Acta,* 53, 259, 1971.

31. **Golden, C. and Sawicki, E.,** Ultrasonic extraction of total particulate aromatic hydrocarbons (TPAH) from airborne particles at room temperatures, *Int. J. Environ. Anal. Chem.,* 4, 9, 1975.

32. **Taylor, D. G., Manual Coordinator,** Benzo(a)pyrene in particulates: analytical method, in *NIOSH Manual of Analytical Methods,* Vol. 1, 251-1, Measurement Research Branch, National Institute for Occupational Safety and Health, U.S. Department of Health, Education and Welfare, Cincinnati, 1977.

33. **Taylor, D. G., Manual Coordinator,** Total particulate aromatic hydrocarbons (TPAH) in air, in *NIOSH Manual of Analytical Methods,* Vol. 1, 206-1, Physical and Chemical Analysis Branch, National Institute for Occupational Safety and Health, U.S. Department of Health, Education and Welfare, Cincinnati, 1977.

34. **Sawicki, E., Belsky, T., Friedman, R. A., Hyde, D. L., Monkman, J. L., Rasmussen, R. A., Ripperton, L. A., and White, L. D.,** Total particulate aromatic hydrocarbons (TPAH) in air. Ultrasonic extraction method, *Health Lab. Sci.,* 12, 407, 1975.

35. **Cautreels, W. and van Cauwenberghe, K.,** Extraction of organic compounds from airborne particulate matter, *Water Air Soil Pollut.,* 6, 103, 1976.

36. **Monkman, J. L., Moore, G. E., and Katz, M.,** Analysis of polynuclear hydrocarbons in particulate pollutants, *Am. Ind. Hyg. Assoc. J.,* 23, 487, 1962.

37. **Bjørseth, A., Bjørseth, O., and Fjeldstad, P. E.,** Polycyclic aromatic hydrocarbons in the workplace atmosphere. I. Determination in an aluminum reduction plant, *Scand. J. Work Environ. Health,* 4, 212, 1978.

38. **Bjørseth, A., Bjørseth, O., and Fjeldstad, P. E.,** Polycyclic aromatic hydrocarbons in the workplace atmosphere. II. Determination in a coke plant, *Scand. J. Work Environ. Health,* 4, 224, 1978.

39. **Das, B. S. and Thomas, G. H.,** Fluorescence detection in high performance liquid chromatographic determination of polycyclic aromatic hydrocarbons, *Anal. Chem.,* 50, 967, 1978.

40. **Searl, T. D., Cassidy, F. J., King, W. H., and Brown, R. A.,** An analytical method for polynuclear aromatic compounds in coke oven effluents by combined use of gas chromatography and ultraviolet absorption spectrometry, *Anal. Chem.,* 42, 956, 1970.

41. **Lawther, P. J., Commins, B. T., and Waller, R. E.,** A study of the concentrations of polycyclic aromatic hydrocarbons in gas works retort houses, *Br. J. Ind. Med.,* 22, 13, 1965.

42. **Stanley, T. W., Meeker, J. W., and Morgan, M. J.,** Extraction of organic compounds from airborne particulates: effects of various solvents and conditions on the recovery of benzo(a)pyrene, benz(c)acridine and 7-H-benz(de)anthracene-7-one, *Environ. Sci. Technol.,* 1, 927, 1967.

43. **Grosjean, D.,** Solvent extraction and organic carbon determination in atmospheric particulate matter: the organic extraction-organic carbon analyzer (OE-OCA) technique, *Anal. Chem.,* 47, 797, 1975.

44. **Griest, W. H., Caton, J. E., Guerin, M. R., Yeatts, L. B., and Higgins, C. E.,** Extraction and recovery of polycyclic aromatic hydrocarbons from highly sorptive matrices such as fly ash, in *Polynuclear Aromatic Hydrocarbons: Chemistry and Biological Effects*, Bjørseth, A. and Dennis, A. J., Eds., Battelle Press, Columbus, Ohio, 1980, 819.

45. **Ball, W. L., Moore, G. E., Monkman, J. L., and Katz, M.,** An evaluation of microsublimation separation of atmospheric polycyclics, *Am. Ind. Hyg. J.*, 23, 222, 1962.

46. **Schultz, M. J., Orheim, R. M., and Bovee, H. H.,** Simplified method for the determination of benzo(a)pyrene in ambient air, *J. Am. Ind. Hyg. Assoc.*, 34, 404, 1973.

47. **Colmsjö, A. and Stenberg, U.,** Vacuum sublimation of polynuclear aromatic hydrocarbons separated by thin-layer chromatography for detection with Shpol'skii low-temperature fluorescence, *J. Chromatogr.*, 169, 205, 1979.

48. **Matsushita, H.,** Analytical methods for monitoring polycyclic aromatic hydrocarbons in the environment, in *Polycyclic Hydrocarbons and Cancer*, Vol. 1, Gelboin, H. V. and Ts'o, P.O.P., Eds., Academic Press, New York, 1978, 71.

49. **Monkman, J. L., Dubois, L., and Baker, C. J.,** The rapid measurement of polycyclic hydrocarbons in air by microsublimation, *Pure Appl. Chem.*, 24, 731, 1970.

50. **Burchfield, M. P., Green, E. E., Wheeler, R. J., and Billedean, S. M.,** Recent advances in the gas and liquid chromatography of fluorescent compounds. I. A direct gas phase isolation and injection system for the analysis of polynuclear arenes in air particles by gas-liquid chromatography, *J. Chromatogr.*, 99, 697, 1974.

51. **Wauters, E., Sandra, P., and Verzele, M.,** Qualitative and semi-quanitative analysis of the non-polar organic fraction of air particulate matter, *J. Chromatogr.*, 170, 125, 1979.

52. **Fitch, W. F. and Smith, D. H.,** Analysis of adsorption properties and adsorbed species on commercial polymeric carbons, *Environ. Sci. Technol.*, 13, 341, 1979.

53. **Griest, W. H., Yeatts, L. B., Jr., and Caton, J. E.,** Recovery of polycyclic aromatic hydrocarbons sorbed on fly ash for quantitative determination, *Anal. Chem.*, 52, 199, 1980.

54. **Hoffman, D., Rathkamp, G., Brunneman, K. D., and Wynder, E. L.,** Chemical studies on tobacco smoke. XXII. Profile analysis of tobacco smoke, *Sci. Total Environ.*, 12, 157, 1973.

55. **Hoffmann, D. and Wynder, E.,** Short term determination of carcinogenic aromatic hydrocarbons, *Anal. Chem.*, 32, 295, 1960.

56. **Novotny, M., Lee, M. L., and Bartle, K. D.,** Methods for fractionation, analytical separation and identification of polynuclear aromatic hydrocarbons, *J. Chromatogr. Sci.*, 12, 606, 1974.

57. **Grimmer, G.,** Eine Methode zur Bestimmung von 3,4-Benzpyren in Tabakrauchkondensaten, *Beitr. Tabakforsch.*, 1, 107, 1961.

58. **Acheson, M. A., Harrison, R. M., Perry, P., and Wellings, R. A.,** Factors affecting the extraction and analysis of polynuclear aromatic hydrocarbons in water, *Water Res.*, 10, 207, 1976.

59. **Grimmer, G. and Böhnke, H.,** Bestimmung des Gesamtgehaltes aller polycyclischen aromatischen Kohlenwasserstoffe in Luftstaub und Kraftfahrzeugabgas mit der Capillar-Gas-Chromatographie, *Z. Anal. Chem.*, 261, 310, 1972.

60. **Haenni, E. O., Joe, F. L., Howard, J. W., and Leibel, R. L.,** A more sensitive and selective absorption criterion for mineral oil, *J. Assoc. Off. Anal. Chem.*, 45, 59, 1962.

61. **Natusch, D. F. S. and Tomkins, B. A.,** Isolation of polycyclic organic compounds by solvent extraction with dimethyl sulfoxide, *Anal. Chem.*, 50, 1429, 1978.

62. **Robbins, W. K.,** Solvent extraction of polynuclear aromatic hydrocarbons, in *Polynuclear Aromatic Hydrocarbons: Chemistry and Biological Effects*, Bjørseth, A. and Dennis, A. J., Eds., Battelle Press, Columbus, Ohio, 1980, 814.

63. **Rosen, A. A. and Middleton, F. M.,** Identification of petroleum refinery wastes in surface waters, *Anal. Chem.*, 27, 790, 1955.

64. **Later, D. W., Lee, M. L., Bartle, K. D., Kong, R. C., and Vassilaros, D. L.,** Chemical class separation and characterization of organic compounds in synthetic fuels, *Anal. Chem.*, 53, 1612, 1981.

65. **Zitko, V.,** Aromatic hydrocarbons in aquatic fauna, *Bull. Environ. Contam. Toxicol.*, 14, 621, 1975.

66. **Grimmer, G. and Hildebrandt, A.,** Kohlenwasserstoffe in der Umgebung des Menschen. I. Eine Methods zur simultanen Bestimmung von dreizehn Kohlenwasserstoffen, *J. Chromatogr.*, 20, 89, 1965.

67. **Moore, G. E., Thomas, R. S., and Monkman, J. L.,** The routine determination of polycyclic hydrocarbons in airborne pollutants, *J. Chromatogr.*, 26, 456, 1967.

68. **Howard, J. W., Fazio, T., White, R. H., and Klimeck, B. A.,** Extraction and estimation of polycyclic aromatic hydrocarbons in total diet composites, *J. Assoc. Off. Anal. Chem.*, 51, 122, 1968.

69. **Griest, W. H., Kubota, H., and Guerin, M. R.,** Resolution of polynuclear aromatic hydrocarbons by packed column GLC, *Anal. Lett.*, 8, 949, 1975.

70. **Severson, R. F., Snook, M. E., Arrondale, R. F., and Chortyk, O. T.,** Gas chromatographic quantitation of polynuclear aromatic hydrocarbons in tobacco smoke, *Anal. Chem.*, 48, 1866, 1976.

71. **Snook, M. E.**, Gel filtration of methyl-substituted polynuclear aromatic hydrocarbons, *Anal. Chim. Acta*, 81, 423, 1976.

72. **Griest, W. H., Tomkins, B. A., Epler, J. L., and Rao, T. K.**, Characterization of multialkylated polycyclic aromatic hydrocarbons in energy-related materials, in *Polycyclic Aromatic Hydrocarbons*, Jones, P. W. and Leber, P., Eds., Ann Arbor Science Publ., Ann Arbor, Mich., 1979, 395.

73. **Grimmer, B., Naujack, K.-W., and Schneider, D.**, Profile analysis of polycyclic aromatic hydrocarbons by glass capillary gas chromatography in atmospheric particulate matter in the nanogram range collecting 10m³ of air, *Fresenius Z. Anal. Chem.*, 311, 475, 1982.

74. **Lee, M. L., Novotny, M., and Bartle, K. D.**, Gas chromatography/mass spectrometric and nuclear magnetic resonance determination of polynuclear aromatic hydrocarbons in airborne particulates, *Anal. Chem.*, 48, 1566, 1976.

75. **Grimmer, G. and Böhnke, H.**, Anreicherung und gaschromatographische Profil-Analyse der polycyclischen aromatischen Kohlenwasserstoffe in Schmierölen, *Chromatographia*, 9, 30, 1976.

76. **Grimmer, G. and Böhnke, H.**, Gas chromatographic profile analysis of polycyclic aromatic hydrocarbons in lubricating oil and fuel, in *Environmental Carcinogens — Selected Methods of Analysis*, Vol. 3, International Agency for Research on Cancer, Lyon, France, 1979, 155.

77. **Schuetzle, D., Lee, F. S.-C., Prater, T. J., and Tejada, S. B.**, The identification of polynuclear aromatic hydrocarbons in mutagenic fractions of diesel particulate extracts, *Int. J. Environ. Anal. Chem.*, 9, 93, 1981.

78. **Ramdahl, T., Becher, G., and Bjørseth, A.**, Nitrated polycyclic aromatic hydrocarbons in urban air particles, *Environ. Sci. Technol.*, 16, 861, 1982.

79. **Oehme, M., Manø, S., and Stray, H.**, Determination of nitrated polycyclic aromatic hydrocarbons in aerosols using capillary gas chromatography combined with different electron capture detection methods, *H. High Resol. Chromatogr. Commun.*, 5, 417, 1982.

80. **Wise, S. A., Bowie, S. L., Chesler, S. N., Cuthrell, W. F., May, W. E., and Rebbert, R. E.**, Analytical methods for the determination of polycyclic aromatic hydrocarbons in air particulate matter, in *Polynuclear Aromatic Hydrocarbons: Physical and Biological Chemistry*, Cooke, M., Dennis, A. J., and Fisher, G. L., Eds., Battelle, Press, Columbus, Ohio, 1982, 919.

81. **Wise, S. A., Bonnett, W. J., and May, W. E.**, Normal-phase and reverse-phase liquid chromatographic separation of polycyclic aromatic hydrocarbons, in *Polynuclear Aromatic Hydrocarbons: Chemistry and Biological Effects*, Bjørseth, A. and Dennis, A. J., Eds., Battelle Press, Columbus, Ohio, 1980, 791.

82. **Vandemark, F. L. and DiCesare, J. L.**, The application of high resolution preparative liquid chromatography to polycyclic aromatic hydrocarbons, in *Polynuclear Aromatic Hydrocarbons: Physical and Biological Chemistry*, Cooke, M., Dennis, A. H., and Fisher, G. L., Eds., Battelle Press, Columbus, Ohio, 1982, 835.

83. **Ogan, K. and Katz, E.**, Analysis of complex samples by coupled-column chromatography, *Anal. Chem.*, 54, 169, 1982.

84. **Dong, M., Locke, D. C., and Ferrand, E.**, High pressure liquid chromatographic method for routine analysis of major parent polycyclic aromatic hydrocarbons in suspended particulate matter, *Anal. Chem.*, 48, 368, 1976.

85. **Pierce, R. C. and Katz, M.**, Dependency of polynuclear aromatic hydrocarbon content on size distribution of atmospheric aerosols, *Anal. Chem.*, 47, 1743, 1975.

86. **Hellmann, M.**, Zur Veränderung der Fluorescenzintensität polycyclischer Aromaten auf Dünnschichtplatten, *Fresenius Z. Anal. Chem.*, 295, 24, 1979.

87. **Issaq, H. J., Andrews, A. W., Janini, G. M., and Barr, E. W.**, Isolation of stable mutagenic photodecomposition products of benzo(a)pyrene by thin-layer chromatography, *J. Liq. Chromatogr.*, 2, 319, 1979.

88. **Lee, M. L., Vassilaros, D. L., and Later, D. W.**, Capillary column gas chromatography of environmental polycyclic aromatic compounds, *Int. J. Environ. Anal. Chem.*, 11, 251, 1982.

89. **Snook, M. E., Severson, R. F., Higman, H. C., Arrendale, R. F., and Chartyk, O. T.**, Polynuclear aromatic hydrocarbons of tobacco smoke: isolation and identification, *Beitr. Tabakforsch.*, 8, 250, 1976.

90. **Grimmer, G., Böhnke, H., and Glaser, A.**, Investigation on the carcinogenic burden by air pollution in man. XV. Polycyclic aromatic hydrocarbons in automobile exhaust gas — an inventory, *Zentralb. Bakteriol. Paraistenkd. Infektionskr. Hyg. Abt. 1, Orig. Reihe B*, 164, 218, 1977.

91. **Bjørseth, A. and Eklund, G.**, Analysis for polynuclear aromatic hydrocarbons in working atmospheres by computerized gas chromatography — mass spectrometry, *Anal. Chim. Acta*, 105, 199, 1979.

92. **Sawicki, E., Stanley, T. W., Elbert, W. C., Meeker, J., and McPherson, S.**, Comparison of methods for determination of benzo(a)pyrene in particulates from urban and other atmospheres, *Atmos. Environ.*, 1, 181, 1967.

93. **Bartle, K. D., Lee, M. L., and Wise, S. A.**, Modern analytical methods for environmental polycyclic aromatic compounds, *Chem. Soc. Rev.*, 10, 113, 1981.

94. **Lee, M. L., Novotny, M. V., and Bartle, K. D.**, *Analytical Chemistry of Polycyclic Aromatic Compounds*, Academic Press, New York, 1981.

95. **Bjørseth, A., Ed.**, *Handbook of Polycyclic Aromatic Hydrocarbons*, Marcel Dekker, New York, 1983.

96. **Gunther, F. A. and Buzzetti, F.**, Occurrence, isolation, and identification of polynuclear hydrocarbons as residue, *Residue Rev.*, 9, 90, 1964.

97. **Sawicki, E., Stanley, T. W., Elbert, W. C., and Pfaff, J. D.**, Application of thin layer chromatography to analysis of atmospheric pollutants and detection of benzo(a)pyrene, *Anal. Chem.*, 36, 497, 1964.

98. **Sawicki, C. R. and Sawicki, E.**, Thin-layer chromatography in air pollution research, in *Progress in Thin-Layer Chromatography and Related Methods*, Vol. 3, Niederwieser, A. and Pataki, G., Eds., Ann Arbor Science Publ., Ann Arbor, Mich., 1972, 233.

99. **Daisey, J. M.**, Analysis of polycyclic aromatic hydrocarbons by thin-layer chromatography, in *Handbook of Polycyclic Aromatic Hydrocarbons*, Bjørseth, A., Ed., Marcel Dekker, New York, 1983, 397.

100. **Bertsch, W., Hara, S., Kaiser, R. E., and Zlatkis, A., Eds.**, *Instrumental HPTLC*, Dr. Alfred Hfhig Verlag, Heidleberg, 1980.

101. **Tomingas, R. and Grover, Y. P.**, Rapid determination of four polycyclic aromatic hydrocarbons by HPLC using nano plates, *Fresenius Z. Anal. Chem.*, 315, 515, 1983.

102. **Wise, S. A.**, High-performance liquid chromatography for the determination of polycyclic aromatic hydrocarbons, in *Handbook of Polycyclic Aromatic Hydrocarbons*, Bjørseth, A., Ed., Marcel Dekker, New York, 1983, 183.

103. **Ogan, K. and Katz, E.**, Retention characteristics of several bonded-phase liquid chromatography columns for some polycyclic aromatic hydrocarbons, *J. Chromatogr.*, 188, 115, 1980.

104. **Katz, E. and Ogan, K.**, Selectivity factors for several PAH pairs on C_{18} bonded phase columns, *J. Liq. Chromatogr.*, 3, 1151, 1980.

105. **Colmsjö, A. L. and Mac-Donald, J. C.**, Column-induced selectivity in separation of polynuclear aromatic hydrocarbons by reversed-phase, high-performance liquid chromatography, *Chromatographia*, 13, 350, 1980.

106. **Amos, R.**, Evaluation of bonded phases for the high-performance liquid chromatographic determination of polycyclic aromatic hydrocarbons in effluent waters, *J. Chromatogr.*, 204, 469, 1981.

107. **Wise, S. A. and May, W. E.**, Effect of C_{18} surface coverage on selectivity in reversed-phase liquid chromatography of polycyclic aromatic hydrocarbons, *Anal. Chem.*, 55, 1479, 1983.

108. **Katz, E. and Ogan, K.**, The effect of mobile phase strength on the selectivity factors for several PAH pairs of different C_{18} columns, *Chromatogr. Newsl.*, 8, 20, 1980.

109. **Chmielowiec, J. and Sawatzky, H.**, Entropy dominated high performance liquid chromatographic separations of polynuclear aromatic hydrocarbons. Temperature as a separations parameter, *J. Chromatogr. Sci.*, 17, 245, 1979.

110. **Lankmayr, E. P. and Müller, K.**, Polycyclic aromatic hydrocarbons in the environment: high-performance liquid chromatography using chemically modified columns, *J. Chromatogr.*, 170, 139, 1979.

111. **Chmielowiec, J. and George, A. E.**, Polar bonded phase sorbents for high performance liquid chromatographic separations of polycyclic aromatic hydrocarbons, *Anal. Chem.*, 52, 1154, 1980.

112. **Sonnefeld, W. J., Zoller, W. H., Wise, S. A., and May, W. E.**, On-line multidimensional liquid chromatographic determination of polynuclear aromatic hydrocarbons in complex samples, *Anal. Chem.*, 54, 723, 1982.

113. **Peaden, P. A., Lee, M. L., Hirata, Y., and Novotny, M.**, High-performance liquid chromatographic separation of high-molecular-weight polycyclic aromatic compounds in carbon black, *Anal. Chem.*, 52, 2268, 1980.

114. **Scott, R. P. W. and Kucera, P.**, Mode of operation and performance characteristics of microbore columns for use in liquid chromatography, *J. Chromatogr.*, 169, 51, 1979.

115. **Knox, J. H. and Gilbert, M. T.**, Kinetic optimization of straight open-tubular liquid chromatography, *J. Chromatogr.* 185, 405, 1980.

116. **Hirose, A., Wiesler, D., and Novotny, M.**, High-efficiency microcolumn separation of polycyclic arene isomers isolated from carbon black, *Chromatographia*, 18, 239, 1984.

117. **Hirata, Y. and Novotny, M.**, Techniques of capillary liquid chromatography, *J. Chromatogr.*, 186, 521, 1979.

118. **Novotny, M., Springston, S. R., Peaden, P. A., Fjeldstad, J. C., and Lee, M. L.**, Capillary supercritical fluid chromatography, *Anal. Chem.*, 53, 407A, 1981.

119. **Gere, D. R., Board, R., and McManigill, D.**, Supercritical fluid chromatography with small particle diameter packed columns, *Anal. Chem.*, 54, 736, 1982.

120. **Peaden, P. A., Lee, M. L., Fjeldstad, J. C., Springston, S. R., and Novotny, M.**, Instrumental aspects of capillary supercritical fluid chromatography, *Anal. Chem.*, 54, 1090, 1982.

121. **Krustlovic, A. M., Rosie, D. M., and Brown, P. R.**, Selective monitoring of polynuclear aromatic hydrocarbons by high pressure liquid chromatography with a variable wavelength detector, *Anal. Chem.*, 48, 1383, 1976.

122. **Fechner, D. and Seifert, B.**, Bestimmung von polycyclischen aromatischen Kohlenwasserstoffen in Staub-niederschlägen durch Hochleistungs-Flüssigkeits-Chromatographie mit Mehrwellenlängendetektion. I. Mitteilung: qualitative Ergebnisse, *Fresenius Z. Anal. Chem.*, 292, 193, 1978.

123. **Sorrell, R. K. and Reding, R.**, Analysis of polynuclear aromatic hydrocarbons in environmental waters by high-pressure liquid chromatography, *J. Chromatogr.*, 185, 655, 1979.

124. **May, W. E. and Wise, S. A.**, Liquid chromatographic determination of polycyclic aromatic hydrocarbons in air particulate extracts, *Anal. Chem.*, 56, 225, 1984.

125. **Jurgensen, A., Inman, E. L., Jr., and Winefordner, J. D.**, Comprehensive analytical figures of merit for fluorimetry of polynuclear aromatic hydrocarbons, *Anal. Chim. Acta*, 131, 187, 1981.

126. **Ogan, K., Katz, E., and Porro, T. J.**, The role of spectral selectivity in fluoresence detection for liquid chromatography, *J. Chromatogr. Sci.*, 17, 597, 1979.

127. **Nielsen, T.**, Determination of polycyclic aromatic hydrocarbons in automobile exhaust by means of high-performance liquid chromatography with fluorescence detection, *J. Chromatogr.*, 170, 147, 1979.

128. **Arpino, P. J. and Guiochon, G.**, LC/MS coupling, *Anal. Chem.*, 51, 682A, 1979.

129. **McFadden, W. H.**, Liquid chromatography/mass spectrometry systems and applications, *J. Chromatogr. Sci.*, 18, 97, 1980.

130. **Schäfer, K. M. and Levsen, K.**, Direct coupling of a micro high-performance liquid chromatograph and a mass spectrometer, *J. Chromatogr.*, 206, 245, 1981.

131. **Dupire, F.**, Gas chromatography at high temperatures. Application to coal tars and their derivatives, *Fresenius Z. Anal. Chem.*, 170, 317, 1959.

132. **Sawicki, E., Fox, F. T., Elbert, W., Hanser, T. R., and Meeker, J.**, Polynuclear aromatic hydrocarbon composition of air polluted by coal tar fumes, *Am. Ind. Hyg. Assoc. J.*, 23, 482, 1962.

133. **Grimmer, G., Hildebrandt, A., and Böhnke, H.**, Investigation on the carcinogenic burden by air pollution in man. II. Sampling and analysis of polycyclic aromatic hydrocarbons in automobile exhaust gas, *Zentralbl. Bakteriol. Parasitenkd. Infektionkr. Hyg. Abt. 1, Orig. Reihe. B.*, 158, 22, 1973.

134. **DeMaio, L. and Corn, M.**, Gas chromatographic analysis of polynuclear aromatic hydrocarbons with packed columns, *Anal. Chem.*, 38, 131, 1966.

135. **Lao, R. C., Thomas, R. S., and Monkman, J. L.**, Computerized GC-MS analysis of PAH in environmental samples, *J. Chromatogr.*, 112, 681, 1975.

136. **Chatot, G., Dangy-Caye, R., and Fontanges, R.**, Etude des hydrocarbures polycycliques de l'atmosphere. II. Application due couplage chromatographie en phase gazeuse-chromatographie sur couches minces au dosage des arènes polynucléaires, *J. Chromatogr.*, 72, 202, 1972.

137. **Burchill, P., Herod, A. A., and James, R. G.**, A comparison of some chromatographic methods for estimation of polynuclear aromatic hydrocarbons in pollutants, in *Carcinogenesis*, Vol. 3, *Polynuclear Aromatic Hydrocarbons*, Jones, P. W. and Freudenthal, R. I., Eds., Raven Press, New York, 1978, 35.

138. **Schutle, K. A., Larsen, D. J., Hornung, R. W., and Crable, J. V.**, Analytical methods used in a study of coke oven effluent, *J. Am. Ind. Hyg. Assoc.*, 36, 181, 1975.

139. **Sauerland, H. D., Stadhelhofer, J., Thomas, R., and Zander, M.**, Neueres aus der Analytik der mehrkernigen Aromaten, *Erdöl. Kohle. Erdgas, Petrochem.*, 30, 215, 1977.

140. **Lao, R. C. and Thomas, R. S.**, The gas chromatographic separation and determination of PAH from industrial processes using glass capillary and packed columns, in *Polynuclear Aromatic Hydrocarbons*, Jones, P. W. and Lever, P., Eds., Ann Arbor Science Publ., Ann Arbor, Mich., 1979, 429.

141. **Snook, M. E., Severson, R. F., Higman, H. C., Arrendale, R. F., and Chortyk, O.T.**, Methods for characterization of complex mixtures of polynuclear aromatic hydrocarbons, in *Polynuclear Aromatic Hydrocarbons*, Jones, P. W. and Leber, P., Eds., Ann Arbor Science Publ., Ann Arbor, Mich., 1979, 231.

142. **Janini, G. M., Johnston, K., and Zielinski, W. L., Jr.**, Gas-liquid chromatographic evaluation and gas chromatography/mass spectrometric application of new high-temperature liquid crystal stationary phases for polycyclic aromatic hydrocarbon separations, *Anal. Chem.*, 48, 1879, 1976.

143. **Janini, G. M.**, Recent usage of liquid crystal stationary phases in gas chromatography, *Adv. Chromatogr.*, 17, 231, 1979.

144. **Grob, K. and Grob, G.**, Trace analysis on capillary columns. Selected practical applications: insecticides in raw butter extract; aroma head space from liquors, auto exhaust gas, *J. Chromatogr. Sci.*, 8, 635, 1970.

145. **Novotny, M.**, Contemporary capillary gas chromatography, *Anal. Chem.*, 50, 16A, 1978.

146. **Jennings, W., Ed.**, *Gas Chromatography with Glass Capillary Columns*, 2nd ed., Academic Press, New York, 1980.

147. **Schulte, E., Ed.**, *Praxis der Kapillar-Gas-Chromatographie*, Springer Verlag, Heidelberg, 1983.

148. **Onuska, F. E., Wolkoff, A. W., Comba, M. E., Larose, R. H., Novotny, M., and Lee, M. L.**, Gas chromatographic analysis of polynuclear aromatic hydrocarbons in shellfish on short, wall-coated glass capillary columns, *Anal. Lett.*, 9, 451, 1976.

149. **Bjørseth, A.**, Analysis of polycyclic aromatic hydrocarbons in particulate matter by capillary gas chromatography, *Anal. Chim. Acta*, 94, 21, 1977.

150. **Lee, M. L. and Wright, B. W.**, Capillary column gas chromatography of polycyclic aromatic compounds, *J. Chromatogr. Sci.*, 18, 345, 1980.

151. **Sortland, B. and Bjørseth, A.**, Analysis of polycyclic aromatic hydrocarbons by gas chromatography, in *Handbook of Polycyclic Aromatic Hydrocarbons*, Bjørseth, A., Ed., Marcel Dekker, New York, 1983, 257.

152. **Jennings, W.**, Evolution and application of the fused silica column, *J. High Resol. Chromatogr. Chromatogr. Commun.*, 3, 601, 1980.

153. **Grob, K.**, Twenty years of glass capillary columns. An empirical model for their preparation and properties, *J. High Resol. Chromatogr. Chromatogr. Commun.*, 2, 599, 1979.

154. **Wright, B. W. and Lee, M. L.**, Rapid analysis using short capillary columns in gas chromatography, *J. High Resol. Chromatogr. Chromatogr. Commun.*, 3, 352, 1980.

155. **Grob, K. and Grob, G.**, Deactivation of glass capillaries by persilylation. III. Extending the wettability by bonding phenyl groups to the glass surface, *J. High Resol. Chromatogr. Chromatogr. Commun.*, 3, 197, 1980.

156. **Schomburg, G., Behlau, H., Dielmann, R., Husmann, H., and Weeke, F.**, Sampling techniques in capillary gas chromatography, *J. Chromatogr.*, 142, 87, 1977.

157. **Grob, K. and Grob, K., Jr.**, On-column injection onto glass capillary columns, *J. Chromatogr.*, 151, 311, 1978.

158. **Bjørseth, A. and Eklund, G.**, Analysis of polynuclear aromatic hydrocarbons by glass capillary gas chromatography using simultaneous flame ionization and electron capture detection, *J. High Resol. Chromatogr. Chromatogr. Commun.*, 1, 22, 1979.

159. **Josefsson, B.**, Mass spectrometric analysis of polycyclic aromatic hydrocarbons, in *Handbook of Polycyclic Aromatic Hydrocarbons*, Bjørseth, A., Ed., Marcel Dekker, New York, 1983, 301.

160. **Oehme, M.**, Determination of isomeric polycyclic aromatic hydrocarbons in air particulate matter by high-resolution gas chromatography/negative ion/chemical ionization mass spectrometry, *Anal. Chem.*, 55, 2290, 1983.

161. **Hertz, H,. S., May, W. E., Wise, S. A., and Chester, S. N.**, Trace organic analysis, *Anal. Chem.*, 50, 428A, 1978.

162. **Vassilaros, D. L., Kong, R. C., Later, D. W., and Lee, M. L.**, Linear retention index system for polycyclic aromatic compounds. Critical evaluation and additional indices, *J. Chromatogr.*, 252, 1, 1980.

163. **Youden, W. J. and Steiner, E. H.**, *Statistical Manual of the Association of Official Analytical Chemists*, Association of Official Analytical Chemists, Arlington, Va., 1975.

164. **Janssen, T.**, Brief review of investigations on PAH in vehicle exhaust, round robin tests on profile analysis, role of fuel and lubricants, field studies, in *Luftverunreinigungen durch polycyclische aromatische Kohlenwasserstoffe — Erfassung und Bewertung*, VDI-Berichte 358, VDI-Verlag, Düsseldorf, 1980, 69.

165. **Bjørseth, A. and Olufsen, B.**, Result from a Scandinavian Round Robin Test of PAH Analysis (in Norwegian), Nordic PAH Proj. Rep. No. 1, Central Institute for Industrial Research, Oslo, 1978.

Chapter 7

BIOLOGICAL MONITORING OF PAH EXPOSURE

There are extensive data available on the biological effect of PAH in general, and BaP in particular, as well as epidemiological studies and chemical characterizations of workplace atmospheres. However, in spite of all these data, there seems to be no clear understanding of the problems connected with occupational hazards of PAH exposure.

There are several prerequisites for particulate PAH to show any biological effect

1. The particles must be deposited in the respiratory tract.
2. PAH must be eluted from the particles.
3. PAH must be taken up in cells, lymph, or blood.
4. PAH must be metabolized to reactive intermediates.

Therefore, determination of PAH concentrations in workplace atmospheres may not give a good measure of the effective dose of PAH received by the individual worker. Questions along the same lines had already been raised in 1959 by Kreyberg,[1] who stated: "If these two facts are correlated: (i.e.) the enormous amounts of 3:4-benzpyrene in the air and the very moderate excess of lung cancer in the gas works, a serious fallacy is evidently involved... This represents a serious warning against any conclusions as to causative relationships between any substance and lung cancer based upon the mere finding of the substance deposited on paper filtering the air." In many cases, however, in which air sampling is of limited or no value, biological monitoring has been successfully used to obtain information on the uptake of hazardous chemicals in the bodies of exposed workers, and to evaluate the significance of their airborne levels.[2-4]

I. INTAKE, DISTRIBUTION, AND EXCRETION OF PAH

The intake of PAH, their distribution in the organism, and their excretion are dependent on numerous physiological, physical, and chemical factors which have still not been sufficiently elucidated. Nevertheless, it is possible to make limited assumptions based on the results of animal studies conducted with several PAH, particularly BaP, a well-known animal carcinogen.

Absorption of PAH through the lung has particular relevance to occupational exposure to PAH in work atmospheres. PAH with four and more rings, which are particularly suspected of being carcinogenic, are almost exclusively adsorbed to particles when they occur in the atmosphere. Depending on several physical, physiological, and anatomical parameters, particulate PAH may be deposited in the upper (nasopharyngeal area), the middle (tracheobronchial area), or the lower (alveolar compartment) respiratory tract.[5]

Mitchell[6-7] showed that 1 to 2 μm aerosol particles of BaP inhaled by rats are deposited mainly in the upper respiratory tract. BaP is cleared from the lungs and transported to the internal organs in two phases: an initial rapid phase where 50% of the BaP was cleared by approximately 2 hr and a second slower phase that continued for about 2 days after exposure. Metabolites of BaP were found in the lung, liver, and kidney shortly after inhalation exposure. The amount of unmetabolized BaP in the lung was no greater than 20%. The results indicated that a major portion of inhaled BaP was metabolized by the lung tissue prior to clearance from the lung. Furthermore, it was found that a significant amount of BaP was covalently bound to the lung tissue and retained for a much longer time than the soluble metabolites.

Sun et al.[8] found that about 10% of pure BaP aerosol (approximately 0.1 μm in diameter) was deposited in the lung. During the exposure, and within the first 30 min after exposure,

Table 1
DISTRIBUTION AND ELIMINATION
OF RADIOACTIVITY AFTER
INTRATRACHEAL INSTILLATION
OF ^{14}C-BaP INTO RATS[a]

	% Of administered dose after:	
Site	**1 hr**	**24 hr**
Feces	0	28.0
Urine	0	1.2
Stomach	0	0
Intestine	37.0	11.1
Kidney	0	2.4
Liver	1.3	4.3
Lung	43.2	38.6
Other organs	0.08	0.43
% Recovered	81.58	86.03

a   Dose: 25 μg ^{14}C-BaP in 0.3 mℓ H_2O.

Adapted from Kotin, P., Falk, H. L., and Busser, R.,
J. Natl. Cancer Inst., 23, 541, 1959.

about 90% of the BaP was cleared from the lung by absorption into blood followed by excretion from the body through the feces. The retention of BaP, however, was much longer when BaP coated on insoluble particles was inhaled.

Furthermore, both studies indicate that a substantial amount of inhaled BaP particles are removed from the lung by mucociliary clearance and subsequent swallowing of the material. This mechanism can result in high PAH levels in the GI tract from which they may be reabsorbed. This may be a second important route of exposure to PAH in work atmospheres.

A third mechanism of absorption of PAH in the organism is by direct contact with the skin. As PAH are highly lipophilic compounds, they are expected to dissolve readily in the lipoprotein membranes and therefore to be rapidly adsorbed by the skin. In a study by Heidelberger and Weiss,[3] topical application of a BaP solution to shaved backs of mice was followed by a diphasic disappearance of radioactivity, with half-lives of 40 and 104 hr for the initial rapid phase and the subsequent slow phase, respectively. Since essentially all of the radioactivity was recovered in the feces within 16 days, quantitative percutaneous absorption of BaP is apparent.

PAH, once adsorbed, becomes localized in a wide variety of body tissues. The distribution of radioactivity derived from ^{14}C-BaP in the rat and mouse was determined following subcutaneous, intravenous, and intratracheal administration.[10] The pattern of distribution was found to be similar in all cases, except for high local pulmonary concentrations following intratracheal instillation (Table 1). Concentrations of BaP-derived radioactivity in the liver reached a maximum within only 10 min after injection and represented 12% of the total dose. Radioactivity in the liver was reduced to 1 to 3% of the administered dose within 24 hr. Similarly, maximum blood levels of BaP following i.v. injection were reached very quickly, and radioactivity became barely detectable after 10 min. Minimal tissue localization of BaP and/or its metabolites occurred in the spleen, kidney, lung, and stomach: maximum radioactivity derived from labeled BaP was recovered in the bile and feces. Levels of radioactivity in fat, skin, and muscle were not determined, nor was the amount of unchanged BaP measured in any tissue. Bock and Dao[11] later showed that relative to other tissues, unmetabolized BaP was located extensively in the mammary gland and general body fat after a single feeding of the carcinogen (10 to 30 mg). This accumulation of BaP was greater than that resulting from 3-methylcholanthrene, 7,12-dimethylbenz(a)anthracene, or phenanthrene. In all cases, the level of carcinogens detected in the tissue was directly related to the dose administered, and was dependent upon the use of a lipid vehicle.

In summary, the results of the studies indicate the following

1. Detectable levels of PAH and/or PAH metabolites can be observed in most internal organs from minutes to hours after administration.
2. Mammary and other fat tissues are significant storage depots where PAH may accumulate and be slowly released.
3. The gut contains relatively high levels of PAH and/or PAH metabolites as a result of hepatobiliary excretion or of the ingestion of particulate PAH following mucociliary clearance after inhalation.

Hepatobiliary excretion and elimination through the feces is the major route by which PAH are removed from the body. Kotin et al.[10] observed that 4 to 12% of the subcutaneously injected dose of BaP was eliminated in the urine of mice within 6 days after injection, while 70 to 75% of the dose was recovered in the feces. Less than 1% of BaP recovered in the bile was unmetabolized. Camus et al.[12] determined fecal excretion rates of BaP in two different strains of mice following i.p. injection of BaP. The cumulative excretion of BaP in the feces followed an exponential curve with half-lives of 1.2 and 0.6 days for the two different strains. The majority of metabolites in the urine appeared as highly water-soluble conjugates. After enzymatic deconjugation, various oxidated BaP metabolites could be identified.

II. METABOLISM AND ACTIVATION OF PAH

PAH are among the chemical carcinogens which are not chemically highly reactive themselves, but exert their carcinogenic activity through metabolites which are sufficiently reactive to modify cellular macromolecules such as nucleic acids (DNA, RNA) and proteins.[13]

Metabolism is dominated by oxidation through the microsomal mixed-function oxidase (MFO) system, often termed aryl hydrocarbon hydroxylase (AHH), which is most abundant in the liver. This enzyme system has been studied extensively and is the subject of several reviews.[14-15] While it is known that this enzyme complex is involved with detoxification of xenobiotics in conjunction with various P-450-type cytochromes, it is apparent that this system is also responsible for the metabolism of polycyclic hydrocarbons to their active species (Figure 1).

The first step in this metabolic activation of PAH, catalyzed by the cytochrome P-450 monooxygenase system, gives rise to epoxide and phenolic groups in different positions on the polycyclic ring system.[16] A second microsomal enzyme, epoxide hydrolase (EH), converts epoxides into vicinal diols.

Information on EH has been summarized recently,[17] and its importance in the formation of three known dihydrodiols of BaP has been demonstrated.[18] Dihydrodiols may be further oxidized by the MFO system to dihydrodiol epoxides.[19] There is now considerable evidence that a particular structural class of diol-epoxides, namely the bay-region dihydrodiol-epoxides, operate as the ultimate carcinogenic form of PAH, which react easily with cellular macromolecules, in particular DNA.[20,21]

Other reactions involved in the metabolism of PAH are enzymatic conjugations of the oxygenated intermediates to glucoronic acid, sulfate, and glutathione.[22] These water-soluble conjugates are readily removed from the organism through bile, feces, and urine, and have been generally viewed as detoxification products.

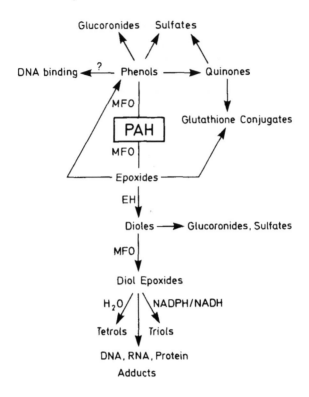

MFO, Multifunctional monooxygenase
EH, Epoxide hydrolase

FIGURE 1. Enzymatic pathways involved in the activation and detoxification of PAH.

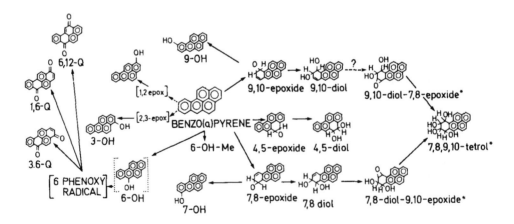

FIGURE 2. Oxidative metabolism of BaP. * Stereochemistry not indicated.

Identification of metabolites has been performed for some PAH, with BaP being the compound studied most extensively. The metabolism of BaP is outlined in Figure 2. BaP metabolites found in microsomal incubation are 1-hydroxy-BaP, 3-hydroxy-BaP, 6-hydroxy-BaP, 7-hydroxy-BaP, and 9-hydroxy-BaP. The BaP-4,5-epoxides have been isolated and identified as precursors of the BaP-4,5-diol. Other studies indicate that epoxides are precursors of the 7,8-diol and 9,10-diol as well. There have been no intermediates isolated as phenol precursors, although recent evidence using deuterium labeling suggests that at least a portion of 3-OH-BaP is derived from an intermediate 2,3-epoxide.[23] In addition to the hydroxylated metabolites, 1,6-, 3,6-, and 6,12-BaP-quinone have been identified.[24] These are produced enzymatically by microsomes and nonenzymatically by air oxidation of phenols.[25]

Further oxidation of 7,8-dihydroxy-7,8-dihydrobenzo(a)pyrene by the MFO system leads to the formation of the highly reactive and probably ultimate carcinogenic metabolites, the isomeric 7,8-dihydroxy-9,10-oxy-7,8,9,10-tetrahydrobenzo(a)pyrene in which the epoxy-ring is adjacent to the "bay" of the hydrocarbon.[26] These diol-epoxides react rapidly by electrophilic attack with cellular macromolecules. Adduct formation has been observed with DNA, RNA, and proteins.[21,27]

The metabolic profile of PAH has been largely worked out using HPLC.[18,28] The reverse-phase technique with aqueous mobile phase and C_{18} modified silica has been used to separate phenols, quinones, and diols from unconverted PAH.[28] The same technique has been applied to the analysis of highly polar metabolites resulting from further metabolism of diols to diol-epoxides. The advantage of HPLC is that it permits collection of individual fractions for further identification by other techniques.

GC separation of PAH metabolites has been described in several reports.[29-31] The enrichment of metabolites and unconverted PAH can be achieved by extraction with ethylacetate, and subsequent chromatography on Sephadex® LH-20. Silylation results in trimethylsilyl ethers which can readily be separated by CGC.[31] Advantages of this method over HPLC are the much better chromatographic resolution, and the possibility of easily combining GC with mass spectrometry for identification and quantitation of the separated metabolites.

III. EFFECTS OF PARTICULATE PAH

Because of low vapor pressures, most carcinogenic PAH in working environments are adsorbed onto particles. It is now well recognized that the nature of the particle may influence the effect of adsorbed PAH. A number of studies have been published on the ability of particulates to enhance or reduce the cellular uptake of PAH. Synergistic effects between particles and PAH have been observed in studies with experimental animals. In the absence of particulate matter, intratracheal instillation of BaP results in only a low incidence of lung cancer, even at high dosages.[32] However, a wide variety of particles, when co-instilled with PAH, result in a high tumor yield. These include hematite.[33] asbestos,[34] carbon particles[35] India ink,[36] and aluminum and titanum oxides.[35] Thus, particulates in general, irrespective of chemical composition, appear to act as co-carcinogens. So far, human epidemiological evidence points to asbestos as being the most significant co-carcinogen.

On the other hand, data resulting from animal studies show that the PAH adsorbed on carbon black is not available for cellular uptake. The results of these animal tests are summarized in Table 2. In this extensive series of studies,[37] whole carbon black and the benzene extract of carbon black were tested in laboratory animals. These tests included skin painting, inhalation, feeding and s.c. injection. The tests for the whole carbon black and the benzene extract of the identical carbon black were performed under identical conditions. In each case the animal tests of the whole carbon black gave negative results, while the benzene extract of the carbon black yielded positive results. Carcinogenic materials can be removed from carbon black by benzene extraction; however, the carcinogenic materials adsorbed on the carbon black apparently are not a carcinogenic hazard.

When inducing lung cancer, there seem to be two essential processes:

1. The inhaled carcinogen must be retained by the respiratory tract. Adsorption of PAH on particles that are themselves efficiently retained can increase the total exposure level to PAH.

Table 2
BIOLOGICAL AVAILABILITY OF
CARCINOGENS FROM CARBON
BLACK

Test method	Whole carbon black	Benzene extract of carbon black
Skin painting	Negative	Positive
Inhalation	Negative	Positive
Feeding	Negative	Positive
Injection	Negative	Positive

Adapted from Nau, C. A., Neal, J., Stenbridge, V. and Cooley, R. N., *Arch. Environ. Health*, 4, 598, 1962.

2.   The carcinogens must enter the cells and be subsequently transformed. Transport can be facilitated by particles that release adsorbed carcinogens rapidly in comparison with the clearance rates of these particles from the lungs, by particles that are rapidly phagocytized, or by particles which penetrate cell membranes as a result of their shapes or surface properties.

Particles that retain the carcinogenic material may possibly decrease the carcinogenic activity. In this regard, Creasia and co-workers[30] showed that when BaP is adsorbed on large carbon particles (15 to 30 μm) and instilled into the lungs, 50% of both the BaP and the carrier particles were cleared from the lungs in 4 to 5 days. Little carcinogenic material was released from the carbon particles in this case, and therefore, contact with the respiratory epithelium (and carcinogenicity) was low. With smaller carbon particles (0.5 to 9.0 μm), however, 50% particle clearance was not achieved until 7 days after administration. In this case, 15% of the adsorbed BaP was eluted from the particles and left free to react with the respiratory tissues. In the complete absence of carrier particles, BaP was cleared from the lungs at 20 times the rate for adsorbed BaP. This observation may explain the difficulty in producing experimental pulmonary tumors with BaP without the use of carrier particles. Other investigators[39] confirmed that carbon particle size affects BaP retention in the lungs, but also demonstrated that BaP retention was not affected by particle size when BaP was adsorbed on ferric oxide or aluminum oxide. Recently, Sun et al.[8] compared the deposition, retention, and biological fate of inhaled BaP as a pure aerosol and BaP adsorbed onto ultrafine particles of Ga_2O_3. Particle adsorption significantly increased the retention of the BaP in the respiratory tract. The results indicate that the rate-limiting step of pulmonary clearance of particulate BaP may be due in part to the slow removal of the BaP coating from the insoluble particles prior to clearance from the lung.

IV. ANALYSIS OF PAH IN URINE

Data on urine levels of various industrial contaminants frequently have been used to supplement information on the concentration of pollutants in the air to which workers are exposed.[2-4] However, data on the analysis of PAH in urine samples are rather limited.

Low concentrations of fluorescent material, presumably unmetabolized PAH, have been observed in urine samples from tobacco smokers and nonsmokers (Table 3).[40,41] Maly[40] used acid hydrolysis of the urine samples petrolether extraction, and paper chromatography for the semiquantitative determination of dibenzo(a,1)pyrene. Repetto and Martinez[41] separated BaP from methylenchloride extracts of urine samples by preparative column chromatography on silica, and quantitated BaP by spectrofluorimetry. Szyja[42] found high amounts of un-

Table 3
CONCENTRATION OF PAH IN HUMAN URINE SAMPLES

PAH	Subjects	No. of samples	Concentration (mean in $\mu g/\ell$)	Ref.
Dibenzo(a,1)pyrene	Active smoker	1	1.1	40
	Passive smoker	1	0.3	40
BaP	Active smoker (morning)	8	0.17	41
	Active smoker (evening)	7	0.34	41
	Passive smoker (morning)	1	0.15	41
	Passive smoker (evening)	1	0.23	41
	Nonexposed (morning)	1	0.01	41
	Nonexposed (evening)	1	0.06	41
BaP	Topside coke oven workers			
	After 6 hr work	19	4.67	42
	After 18 hr rest	13	2.1	42
	After 48 hr rest	8	1.6	42
BaP	Persons from urban/industrial area	451	0.690	43
	Persons from rural area	35	0.445	43
Benz(a)anthracene	Persons from urban/industrial area	437	0.701	43
	Persons from rural area	34	0.489	43
Sum of 11 prominent PAH[a]	Aluminum workers			
	Smokers	7	68.2[b]	46
	Nonsmokers	4	79.3[b]	46
	Controls			
	Smokers	6	45.4[b]	46
	Nonsmokers	4	13.2[b]	46
Sum of 10 prominent PAH[a]	Aluminum workers			47,48
	Smokers	9	47.5[b]	47,48
	Nonsmokers	6	24.2[b]	
	Controls			
	Smokers	4	43.4[b]	47,48
	Nonsmokers	5	6.3[b]	47,48

[a] Sum of PAH and PAH metabolites determined by reversed-metabolism method.
[b] Results converted from μg PAH/mmol creatinine using an average excretion of 12 mmol creatinine per 1 urine.

metabolized BaP in urine samples of topside coke oven workers, collected after 6 hr work (Table 3). The urinary PAH levels were significantly lower after 18- and 48-hr rests. Michels and Einbrodt[43] determined the concentration of BaP and benz(a)anthracene (BaA) in more than 480 urine samples of randomly selected persons from a highly industrialized area, and from a rural area as reference. The excretion of both BaP and BaA was found to be significantly higher in the polluted area compared to the reference area (Table 3). No differences however, were observed between urines from smokers and nonsmokers.

In the studies described above, only the unmetabolized part of the PAH excreted in urine was determined. However, as mentioned earlier, PAH are metabolized to a great extent to polar, water-soluble metabolites both in vitro and in vivo. For example, the relative amount of unmetabolized BaP found in urine from mice following i.p. injection of ^{14}C-labeled BaP was as low as 0.7% of the total amount excreted in urine.[44] Thus, an alternative approach has been suggested by Keimig et al.[45] to analyze urine for metabolites of specific PAH

Table 4
PAH IN AN EXPOSED WORKER'S URINE WITHOUT AND
WITH REDUCTION OF METABOLITES

PAH	Without reduction (unmetabolized PAH)	With reduction (unmetab. + metabol. PAH)
Fluorene	n.d.[a]	4.2
Phenanthrene	6.85	9.8
Anthracene	n.d.[a]	4.4
Fluoranthene	1.12	9.8
Pyrene	0.37	5.5
Benzo(a)fluorene	n.d.[a]	3.1
BaA	0.17	2.0
Chrysene	0.18	Obscured
BeP	n.d.[a]	0.7
BaP	0.01	0.13
Dibenz(a,h)anthracene	n.d.[a]	0.57
Sum	8.7	40.2

Note: Concentrations are in $\mu g/\ell$.

[a] n.d. = not detected.

Adapted from Becher, G. and Bjørseth, A., *Cancer Lett.*, 17, 301, 1983.

compounds that are consistently prominent in environmental samples. They have identified 1-hydroxypyrene as a major metabolite in the urine of pigs and propose this metabolite for monitoring PAH-exposed workers. However, no application of this technique in occupational hygiene has been reported so far. Recently, Becher and Bjørseth[46] have described a method to determine multiple PAH compounds in urine specimens based on the reduction of excreted, oxidated metabolites of PAH back for the parent hydrocarbons. The analytical procedure included extraction of PAH and PAH metabolites from urine using cartridges containing C_{18} modified silica, reduction of metabolites to PAH by refluxing hydriodic acid ("reversed metabolism"), and subsequent analysis of 11 prominent PAH by HPLC with fluorimetric detection. Table 4 shows the results for two parallel analyses, with one including the reduction of metabolites and the other one omitting the reduction step. The table reveals that the amount of PAH identified increased approximately fivefold when the reduction step is included. The method has been applied to urine samples from aluminum workers with high exposure to PAH and to occupationally nonexposed control groups. Figure 3 shows a typical HPLC/fluorescence chromatogram of PAH isolated from an exposed worker's urine. The results are included in Table 4. In the control group, urine extracts from smokers show a significantly higher level of PAH than from nonsmokers. However, the high concentrations of PAH found in the working atmospheres of aluminum plants are not reflected to a corresponding extent in the excretion of PAH in workers' urine.

In a more recent study, simultaneous monitoring of PAH exposure of the aluminum workers was performed using personal sampling of airborne particulate PAH as well as urine analysis.[47,48] The average particulate PAH exposure was 126 $\mu g/m^3$. The PAH profile in urine seems to correspond with the profile found in the work atmosphere, if one takes into account the sampling efficiency of the filter used for sampling particulates is decreasing with increasing volatility of the PAH. The mean of PAH concentrations in the urine samples from the four different categories of personnel investigated are given in Table 3 and shown schematically in Figure 4. A significant increase in the excretion of PAH and PAH-metabolites is observed for smokers in the control groups as compared to the nonsmokers in this

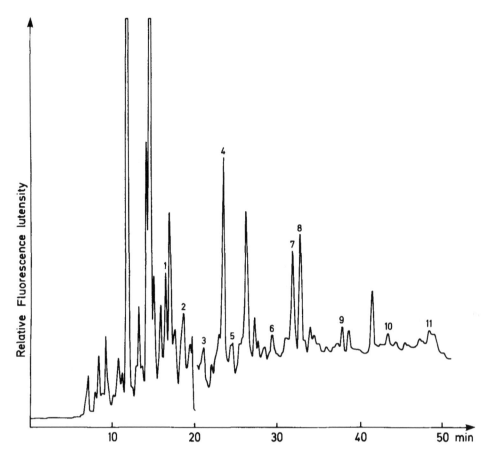

FIGURE 3. Typical reverse-phase HPLC chromatogram of PAH in urine from an exposed worker after reduction of metabolites. Peak identities: 1, fluorene; 2, phenanthrene; 3, anthracene; 4, fluoranthene; 5, pyrene; 6, benzo(a)fluorene; 7, BaA; 8, chrysene; 9, BeP; 10, BaP; 11, dibenz(a,h)anthracene. (From Becher, G. and Bjørseth A., *Cancer Lett.*, 17, 301, 1983. With permission.)

group. However, in accordance with previous results, the high occupational PAH exposure is not reflected to a great extent in the excretion of PAH in the urine.

These results are particularly interesting when comparing the analytical data with epidemiological data on occupational cancer in the aluminum industry. There is no obvious correlation between exposure, as measured by air analysis, and lung cancer frequency. Recent epidemiological data indicate that there is only a slight excess of lung cancer in aluminum workers, while the exposure is up to three orders of magnitude higher compared to urban atmospheres. On the other hand, analyses of PAH metabolites in urine fit well with the epidemiological data, as there is only a slight increase in the excretion of PAH in exposed workers. This corresponds well with the low or negligible increase in lung cancer found in epidemiological studies. The results are supported by the study of sister chromatid exchange (SCE) in blood lymphocytes of the same aluminum workers.[48] The highly exposed workers had no significant increase in the SCE frequencies, while the SCE frequencies of cigarette smokers were significantly higher than those of the nonsmokers in both groups.

V. ANALYSIS OF PAH IN BLOOD SAMPLES

In many cases, blood has been recognized as a useful matrix for biological monitoring, as blood levels often reflect the dose, i.e., the amount of toxic agent in the target organ or tissue.[49]

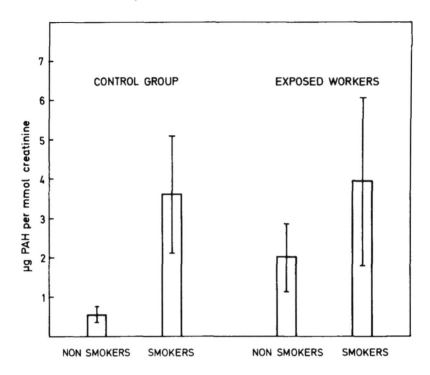

FIGURE 4.   Mean ± SD of PAH levels in urine samples from aluminum workers and controls. (From Becher, G. Haugen, Aa., and Bjørseth, A., *Carcinogenesis*, 5, 647, 1984. With permission.)

It has been shown that PAH form complexes with serum albumin by hydrophobic interaction.[50] This interaction seems to play an important role in the transport of PAH in the blood system. Recently, Hutcheon and co-workers[51] determined plasma BaP levels by radioimmunoassay. They observed a higher mean BaP level in subjects from an urban industrialized area compared to subjects from an outer suburban area.

Recently, extensive interest has been focused on the formation of covalent PAH adducts with cellular macromolecules to evaluate the risks posed by exposure to mutagenic and carcinogenic chemicals.[52-55] The interaction of activated carcinogens with cellular targets (DNA and possibly RNA and proteins) is a critical early event in the action of a number of carcinogens, among the PAH, and is presumed to be directly involved in the carcinogenic process.[54] Therefore, it seems reasonable to assume that the quantitation of PAH-DNA adducts can be used as a particularly relevant measure of the biologically effective dose of the metabolically activated PAH. Furthermore, the carcinogenic potency of a number of PAH correlates with their ability to form covalent adducts with DNA.[53] Ultrasensitive immunoassays specific for BaP-DNA adducts have been developed[56,57] which allow the detection of one BaP-DNA adduct per 10^6 nucleotides. The methods could be sensitive enough to detect adduct formation in vivo as a result of environmental exposure.

In a pilot project, Perera et al.[57] observed low levels of BaP-DNA adducts in lung tissues from lung cancer patients. However, the number of subjects was too small to draw conclusions relating exposure history to the occurrence of PAH-DNA adducts. Monitoring the biological effective dose of PAH in exposed workers restricts the type of sample as a source for DNA to blood. Although some qualitative results from the adduct level of DNA from peripheral blood mononuclear cells are reported by Perera et al., it is not quite clear if the assay method used is sensitive enough to detect the low levels in blood cells.

Fluorimetry, with its intrinsically high sensitivity, has been used as an alternative to immunoassays for determining the binding of PAH to DNA.[58] Fluorescence studies of intact BaP-DNA adducts have utilized either photon counting,[59] low-temperature fluorimetry,[60] or a combination of both of these methods[61] to detect as little as one molecule of bound BaP per 10^5 DNA nucleotides. Recently, Rahn et al.[62,63] described a fluorimetric HPLC assay based on the acid hydrolysis of DNA to liberate the BaP adducts in the form of the isomeric tetrols of BaP, which are quantified by normal fluorescence detection after HPLC separation.

Until now, only limited data have been published on the application of the fluorimetric methods to DNA samples from humans.[64] However, extensive research in this field is presently going on in various research groups.

Measurements of the extent of adduct formation with hemoglobin in blood samples taken from exposed individuals have been proposed as markers of the biological effective dose.[52,65] This approach, called "hemoglobin dosimetry", is based on the fact that most activated carcinogens bind covalently not only to nucleic acid but also to proteins. In fact, the reactivity per unit weight of hemoglobin towards activated carcinogens is about 1.5 to 10 times higher than that of DNA. Thus, hemoglobin serves as a convenient trapping agent for assessing the levels of activated carcinogens in vivo since it can be obtained in large amounts from RBC and is readily purified and analyzed. Furthermore, human erythrocytes have an average life span of about 125 days. Assuming reaction with the activated carcinogen does not have a major influence on the lifetime of the erythrocytes, they will integrate the dose over a long time.

Data from experiments with mice show that a body dose of 1 μg BaP per kg mouse gives 0.08 to 0.10 μg BaP bound to 1 g hemoglobin.[66] Taking into account the accumulation of adducts during the lifetime of the erythrocytes, this degree of binding should be sufficient to be measured by modern analytical methods.

VI. SUMMARY AND CONCLUSION

The biological effect of airborne PAH depends on several factors such as inhalability, deposition in the respiratory tract, bioavailability, and metabolism. Therefore, the mere determination of PAH concentrations in workplace atmospheres may be questioned as a measure of the occupational hazard connected with PAH exposure. It is suggested that analysis of body fluids be used to determine the uptake of PAH in the organism.

There are two approaches to measuring the amount of PAH that has actually entered the body:

1. Detection and quantitation of PAH and PAH-metabolites in the body fluid
2. Measurement of the extent of adduct formation of PAH with cellular macromolecules

A method has been developed to determine the PAH and PAH metabolites in human urine. The amount of PAH found in urine is a relative measure of the total amount of PAH adsorbed in the organism, metabolized, and excreted.

Highly sensitive analytical methods are available for detecting the levels of adducts of PAH and DNA or hemoglobin isolated from blood samples of exposed workers. These will give a measure of the biological effective dose of PAH at the cellular level under actual conditions of worker exposure, i.e., the amount of the metabolically activated PAH that has reacted with cellular targets. Hitherto, only a few data have been published on the application of the adduct methods to samples from humans.

REFERENCES

1. **Kreyberg, L.,** 3:4-Benzpyrene in industrial air pollution: some reflections, *Br. J. Cancer,* 13, 618, 1959.
2. **Linch, A. L.,** *Biological Monitoring for Industrial Chemical Exposure Control.* CRC Press, Boca Raton, Fla., 1974.
3. **Baselt, R. C.,** *Biological Monitoring Methods for Industrial Chemicals,* Biomedical Publ., Davis, Calif., 1980, 301.
4. **Lauwerys, R.,** Biological criteria for selected industrial toxic chemicals: a review, *Scand. J. Work Environ. Health,* 1, 139, 1975.
5. International Radiological Protection Commission, Task Group on Lung Dynamics, Deposition of retention models for internal dosimetry of the human respiratory tract, *Health Phys.,* 12, 173, 1966.
6. **Mitchell, C. E.,** Distribution and retention of benzo(a)pyrene in rats after inhalation, *Toxicol. Lett.,* 11, 35, 1982.
7. **Mitchell, C. E.,** The metabolic fate of benzo(a)pyrene in rats after inhalation, *Toxicology,* 28, 65, 1983.
8. **Sun, J. D., Wolff, R. K., and Kanapilly, G. M.,** Depostion, retention, and biological fate of inhaled benzo(a)pyrene adsorbed onto ultrafine particles and as a pure aerosol, *Toxicol. Appl. Pharmacol.,* 65, 231, 1982.
9. **Heidelberger, C. and Weiss, S. M.,** The distribution of radioactivity in mice following administration of 3,4-benzo-pyrene-5-C^{14} and 1,2,5,6-dibenzanthracene-9,10-C^{14}, *Cancer Res.,* 11, 885, 1951.
10. **Kotin, P., Falk, H. L., and Busser, R.,** Distribution, retention, and elimination of C^{14}-3,4-benzpyrene after administration to mice and rats, *J. Natl. Cancer Inst.,* 23, 541, 1959.
11. **Bock, F. G. and Dao, T. L.,** Factors affecting the polynuclear hydrocarbon level in rat mammary glands, *Cancer Res.,* 21, 1024, 1961.
12. **Camus, A.-M., Aitio, A., Sabadie, N., Wahrendorf, J., and Bartsch, H.,** Metabolism and urinary excretion of mutagenic metabolites of benzo(a)pyrene in C57 and DBA mice strains, *Carcinogenesis,* 5, 35, 1984.
13. **Miller, E. C. and Miller, J. A.,** Biochemical mechanisms of chemical carcinogenesis, in *The Molecular Biology of Cancer,* Busch, H. M., Ed., Academic Press, New York, 1973, 377.
14. **Ishimura, Y., Iizuka, T., Morishima, I., and Hayaishi, O.,** Enzymes of oxygenation, in *Polycyclic Hydrocarbons and Cancer,* Vol. 1, Gelboin, H. V. and Ts'o, P. O. P., Eds., Academic Press, New York, 1978, 321.
15. **Estabrook, R. W., Werringloer, J., Capdevila, J., and Prough, R. A.,** The role of cytochrome P-450 and the microsomal electron transport system: the oxidate metabolism of benzo(a)pyrene, in *Polycyclic Hydrocarbons and Cancer,* Vol. 1, Gelboin, H. V. and Ts'o, P. O. P., Eds., Academic Press, New York, 1978, 285.
16. **Sims, P. and Grover, P. L.,** Epoxides in polycyclic aromatic hydrocarbon metabolism and carcinogenesis, *Adv. Cancer Res.,* 20, 165, 1974.
17. **Guenthner, T. M. and Oesch, F.,** Microsomal epoxide hydrolase and its role in polycyclic aromatic hydrocarbon biotransformation, in *Polycyclic Hydrocarbons and Cancer,* Vol. 3, Gelboin, H. V. and Ts'o, P. O. P., Eds., Academic Press, New York, 1981, 183.
18. **Yang, S. K., Roller, P. P., and Gelboin, H. V.,** Enzymatic mechanism of benzo(a)pyrene conversion to diols and phenols and an improved high-pressure liquid chromatographic separation of benzo(a)pyrene derivatives, *Biochemistry,* 16, 3680, 1977.
19. **Sims, P., Grover, P. L., Swaisland, A., Pal, K., and Hewer, A.,** Metabolic activation of benzo(a)pyrene proceeds by a diol-epoxide, *Nature (London),* 252, 326, 1974.
20. **King, H. W. S., Osborne, M. R., Beland, F. A., Harvey, R. G., and Brookes, P.,** ($\pm$) 7α, 8β-dihydroxy-9β, 10β-epoxy-7,8,9,10-tetrahydrobenzo(a)pyrene as an intermediate in the metabolism and binding to DNA of benzo(a)pyrene, *Proc. Natl. Acad. Sci. U.S.A.,* 73, 2679, 1976.
21. **Phillips, D. H. and Sims, P.,** Polycyclic aromatic hydrocarbon metabolites: their reactions with nucleic acids, in *Chemical Carcinogens and DNA,* Vol. 2, Grover, P. L., Ed., CRC Press, Boca Raton, Fla., 1979, 29.
22. **Nemoto, N.,** Glutathione, glucuronide, and sulfate transferase in polycyclic aromatic hydrocarbon metabolism, in *Polycyclic Hydrocarbons and Cancer,* Vol. 3, Gelboin, H. V. and Ts'o, P. O. P., Eds., Academic Press, New York, 1981, 213.
23. **Yang, S. K., Roller, P. P., Fu, P. P., Harvey, R. G., and Gelboin, H. V.,** Evidence for a 2,3-epoxide as an intermediate in the microsomal metabolism of benzo(a)pyrene to 3-hydroxybenzo(a)pyrene, *Biochem. Biophys. Res. Commun.,* 77, 1176, 1977.
24. **Lesko, S. P., Caspary, W., Lorentzen, R., and Ts'o, P. O. P.,** Enzymatic formation of 6-oxy-benz(a)pyrene radical in rat liver homogenates from carcinogenic benzo(a)pyrene, *Biochemistry,* 14, 3978, 1975.

25. **Lorentzen, R. J., Caspary, W. J., Lesko, S. A., and Ts'o, P. O. P.,** The autoxidation of 6-hydroxy-benzo(a)pyrene and 6-oxobenzo(a)pyrene radical, reactive metabolites of benzo(a)pyrene, *Biochemistry,* 14, 3970, 1975.

26. **Jerina, D. M., Yagi, H., Lehr, R. E., Thakker, D. R., Schaefer-Ridder, M., Karle, J. M., Levin, W., Wood, A. W., Chang, R. L., and Conney, A. H.,** The bay-region theory of carcinogenesis by polycyclic aromatic hydrocarbons, in *Polycyclic Hydrocarbons and Cancer,* Vol. 1, Gelboin, H. V. and Ts'o, P. O. P., Eds., Academic Press, New York, 1978, 173.

27. **Weinstein, I. B., Jeffrey, A. M., Leffler, S., Pulkabek, P., Yamasaki, H., and Grunberger, D.,** Interaction between polycyclic aromatic hydrocargons and cellular macromolecules, in *Polycyclic Hydrocarbons and Cancer,* Vol. 2, Gelboin, H. V. and Ts'o, P. O. P., Eds., Academic Press, New York, 1978, 3.

28. **Selkirk, J. K.,** Analysis of benzo(a)pyrene metabolites by high-pressure liquid chromatography, *Adv. Chromatogr.,* 16, 1, 1978.

29. **Bettencourt, A., Lhoest, G., Roberfroid, M., and Mercier, M.,** Gas chromatographic and mass fragmentographic assays of carcinogenic polycyclic hydrocarbon epoxide hydratase activity, *J. Chromatogr.,* 134, 323, 1977.

30. **Takahashi, G., Kinoshita, K., Hashimoto, K., and Yasuhira, K.,** Identification of benzo(a)pyrene metabolites by gas chromatograph-mass spectrometer, *Cancer Res.,* 39, 1814, 1979.

31. **Jacob, J.,** Analysis of metabolites of polycyclic aromatic hydrocarbons by GC and GC/MS, in *Handbook of Polycyclic Aromatic Hydrocarbons,* Bjørseth, A., Ed., Marcel Dekker, New York, 1983, 617.

32. **Saffiotti, U., Cefis, F., Kolb, L. H., and Shubik, P.,** Experimental studies of the condition of exposure to carcinogens for lung cancer indication, *J. Air Pollut. Control Assoc.,* 15, 23, 1965.

33. **Saffiotti, U., Cefis, F., and Kolb, L. H.,** A method for the experimental indication of bronchogenic carcinoma, *Cancer Res.,* 28, 104, 1968.

34. **Pylev, L. N. and Shabad, K. M.,** Some results of experimental studies in asbestos carcinogenesis, in *Biological Effects of Asbestos,* IARC Sci. Publ. No. 8, Bogovski, P., Timbrell, V., Gilson, J. C., and Wagner, J. C., Eds., International Agency for Research on Cancer, Lyon, France, 1973, 99.

35. **Stenbäck, F., Rowland, J., and Sellakumar, A.,** Carcinogenicity of benzo(a)pyrene and dusts in hamster lungs (instilled intratracheally with titanium oxide, aluminum oxide, carbon and ferric oxide), *Oncology,* 33, 29, 1976.

36. **Pylev, L. N.,** Experimental induction of lung cancer in rats by intratracheal administration of 9,10—dimethyl-1,2-dibenzanthracene, *Bull. Exp. Biol. Med. (U.S.S.R.),* 52, 1316, 1961.

37. **Nau, C. A., Neal, J., Stenbridge, V., and Cooley, R. N.,** Physiological effects of carbon black. IV. Inhalation, *Arch. Environ. Health,* 4, 598, 1962, and previous papers in this series.

38. **Creasia, D. A., Poggenburg, J. K., and Nettesheim, P.,** Elution of benzo(a)pyrene from carbon particles in the respiratory tract of mice, *J. Toxicol. Environ. Health,* 1, 967, 1976.

39. **Henry, M. C. and Kaufman, D. G.,** Clearance of benzo(a)pyrene from hamster lungs after administration on coated particles, *J. Natl. Cancer Inst.,* 51, 1961, 1973.

40. **Maly, E.,** A simple test for exposure to polycyclic hydrocarbons, *Bull. Environ. Contam. Toxicol.,* 6, 422, 1971.

41. **Repetto, M. and Martinez, D.,** Benzopyrene de cigarette et son excrétion urinaire, *J. Eur. Toxicol.,* 7, 234, 1974.

42. **Szyja, J.,** Untersuchungen über den 3,4-Benzpyrengehalt an Arbeitsplätzen der Pechofenbatterie einer Kokerei und in Körperflüssigkeiten der Arbeiter, *Z. Gesamte Hyg. Ihre Grenzgeb.,* 7, 440, 1977.

43. **Michels, S. and Einbrodt, H. J.,** Polycyclic aromatic hydrocarbons in human urines collected in a large industrial city — an epidemiological study (in German), *Wiss. Umwelt.,* 107, 1979.

44. **Becher, G. and Löfroth, G.,** unpublished results.

45. **Keimig, S. D., Kirby, K. W., Morgan, D. P., Keiser, J. F., and Huberg, T. D.,** Identification of 1-hydroxypyrene as a major metabolite in pig urine, *Xenobiotica,* 13, 415, 1983.

46. **Becher, G. and Bjørseth, A.,** Determination of exposure to polycyclic aromatic hydrocarbons by analysis of human urine, *Cancer Lett.,* 17, 301, 1983.

47. **Becher, G. and Bjørseth, A.,** A novel method for the determination of occupational exposure to polycyclic aromatic hydrocarbons by analysis of body fluids, in *Polynuclear Aromatic Hydrocarbons: Mechanisms, Methods and Metabolism,* Cooke, M. and Dennis, A. J., Eds., Batelle Press, Columbus, Ohio, 1985, 145.

48. **Becher, G., Haugen, Aa., and Bjørseth, A.,** Multimethod determination of occupational exposure to polycyclic aromatic hydrocarbons in an aluminum plant, *Carcinogenesis,* 5, 647, 1984.

49. **Grunder, F. I. and Moffitt, A. E.,** Blood as a matrix for biological monitoring, *Am. Ind. Hyg. Assoc. J.,* 43, 271, 1982.

50. **Franke, R.,** Die hydrophobe Wechselwirkung von polycyclischen aromatischen Kohlenwasserstoffen mit Humanserumalbumin, *Biochim. Biophys. Acta,* 160, 378, 1968.

51. **Hutcheon, D. E., Kantrowitz, J., van Gelder, R. N., and Flynn, E.,** Factors affecting plasma benzo(a)pyrene levels in environmental studies, *Environ. Res.,* 32, 104, 1983.

52. **Ehrenberg, L. and Osterman-Golkar, S.,** Alkylation of macromolecules for detecting mutagenic agents, *Teratogenesis Carcinogenesis Mutagenesis,* 1, 105, 1980.

53. **Lutz, W. K.,** In vivo covalent binding of organic chemical to DNA as a quantitative indicator in the process of chemical carcinogenesis, *Mutation Res.,* 65, 289, 1979.

54. **Bridges, B. A.,** An approach to the assessment of the risk to man from DNA damaging agents, *Arch. Toxicol. Suppl.,* 3, 271, 1980.

55. **Perera, F. and Weinstein, I. B.,** Molecular epidemiology and carcinogen-DNA adduct detection: new approaches to studies of human cancer causation, *J. Chronic Dis.,* 35, 581, 1982.

56. **Hsu, I.-C., Poirier, M. C., Yuspa, S. H., Grunberger, D., Weinstein, I. B., Yolken, R. H., and Harris, C. C.,** Measurement of benzo(a)pyrene-DNA adducts by enzyme immunoassays and radioimmunoassay, *Cancer Res.,* 41, 1090, 1981.

57. **Perera, F. P., Poirier, M. C., Yuspa, S. H., Nakayama, J., Jaretzki, A., Curnen, M. M., Knowles, D. M., and Weinstein, I. B.,** A pilot project in molecular cancer epidemiology: determination of benzo(a)pyrene-DNA adducts in animal and human tissues by immunoassays, *Carcinogenesis,* 3, 1405, 1982.

58. **Vigny, P. and Duquesne, M.,** Fluorimetric detection of DNA-carcinogen complexes, in *Chemical Carcinogens and DNA,* Vol. 2, Grover, P. L., Ed., CRC Press, Boca Raton, Fla., 1979, 85.

59. **Daudel, P., Duquesne, M., Vigny, P., Grover, P. L., and Sims, P.,** Fluoresence spectral evidence that benzo(a)pyrene-DNA-products in mouse skin arise from diol-epoxides, *FEBS Lett.,* 57, 250, 1975.

60. **Ivanovic, V., Geactinov, N. E., and Weinstein, I. B.,** Cellular binding of benzo(a)pyrene to DNA characterized by low temperature fluorescence, *Biochem. Biophys. Res. Commun.,* 70, 1172, 1976.

61. **Rahn, R. O., Chang, S. S., Holland, J. M., Stephens, T. J., and Smith, L. H.,** Binding of benzo(a)pyrene to epidermal DNA and RNA as detected by synchronous luminescence spectrometry at 77 K, *J. Biochem. Biophys. Methods,* 3, 285, 1980.

62. **Rahn, R. O., Chang, S. S., Holland, J. M., and Shugart, L. R.,** A fluorometric-HPLC assay for quantitating the binding of benzo(a)pyrene metabolites to DNA, *Biochem. Biophys. Res. Commun.,* 109, 262, 1982.

63. **Shugart, L., Holland, J. M., and Rahn, R. O.,** Dosimetry of PAH skin carcinogenesis: covalent binding of benzo(a)pyrene to mouse epidermal DNA, *Carcinogenesis,* 4, 195, 1983.

64. **Harris, C. C., Vahakangas, K., Newman, M. J., Shamsuddin, A., Sinopoli, N., Mann, D. L., and Wright, W. E.,** Detection of benzo(a)pyrene diole epoxide-DNA adducts in peripheral blood lymphocytes and antibodies to the adduct in serum from coke oven workers, *Proc. Natl. Acad. Sci. U.S.A.,* 82, 6672, 1985.

65. **Calleman, C. J., Ehrenberg, L., Jansson, B., Osterman-Golkar, S., Segerbäck, D., Svensson, K., and Wachtmeister, C. A.,** Monitoring and risk assessment by means of haemoglobin alkylation in persons occupationally exposed to ethylene oxide, *J. Environ. Pathol. Toxicol.,* 2, 427, 1978.

66. **Löfroth, G. and Ehrenberg, L.,** unpublished results.

Chapter 8

PAH IN DIFFERENT WORKPLACE ATMOSPHERES

I. THRESHOLD LIMIT VALUES FOR PAH

There are several factors which make the establishment of threshold limit values (TLVs) for PAH in work atmospheres a difficult task. These include the long latency period between first exposure and the expression of cancer, the lack of knowledge of the actual dose-response relationship for PAH, and the fact that particulate PAH contains a great number of individual compounds with varying carcinogenic activity which can act to enhance or diminish the carcinogenic effect of each other.

Thus, TLVs for PAH were practically nonexistent until 1967, when the American Conference of Governmental Industrial Hygienists (ACGIH) adopted a TLV of 0.2 mg/m^3 for the BSF of the particulates sampled on a filter. As with most occupational limits, it refers to an exposure averaged over an 8-hr shift. In 1976, the U.S. Occupational Safety and Health Administration (OSHA) issued a standard for occupational exposure to coke oven emissions in which a lowering of the TLV to 0.15 mg/m^3 was recommended.[1] The limit was justified as a reasonable level which management could be expected to achieve through engineering controls and good work practices. This lowering of the TLV has not, so far, been legally accepted in the U.S. and the latest edition of the OSHA TLV list still has 0.2 mg/m^3.[2]

NIOSH, in a criteria document,[3] proposed the substitution of benzene with cyclohexane for the extraction of the particulates sampled, since the latter solvent is far less toxic. The permissible exposure limit was recommended to be 0.1 mg/m^3 for the cyclohexane-extractable fraction of the particulates, determined as a time-weighted average concentration for up to a 10-hr work shift in a 40-hr work week. This value is the lowest concentration that can be reliably detected by the recommended method. This TLV still remains a proposal.

Other countries have established TLVs for BaP, which are used as indicators of ambient exposure to carcinogenic PAH in general. In the 1970s, the U.S.S.R. issued a TLV for BaP (0.15 μg/m^3). The documentation of this value, based on experiments with laboratory animals, was given by Shabad.[4] In Sweden, a TLV of 10 μg/m^3 was introduced in 1978. It was lowered to 5 μg/m^3 in 1982.

Recently, an administrative norm of 40 μg/m^3 for occupational exposure to total particulate PAH has been introduced in Norway.[5] It is based on a detailed PAH analysis of particulate samples by GLC.

Thus, in the literature there exist three different types of data on the occurrence of PAH in workplace atmosphere; namely, data on BSF, on BaP alone, and on the sum of individual PAH components.

Most work so far has focused on the gravimetric determination of benzene- or cyclohexane-soluble organic matter.[3] This method has been adopted because it is rather simple and is expected to give a general view of the total exposure to coal tar pitch volatiles. However, as was discussed in detail in Chapter 6, Section II, this method is ambiguous due to a variety of interferences, and does not correlated well with determinations of PAH.

Many studies have been restricted to the determination of BaP because of its well-documented carcinogenicity and the relative ease of analysis. Of course, knowledge of BaP alone is not sufficient for estimating the cancer hazard in a given environment because of the following reasons:

1.   The PAH composition and BaP level can be highly variable depending on the sources

Table 1
EXPOSURE TO BSF OF COKE OVEN PARTICULATES
ACCORDING TO JOB TITLE

Job title	Number of samples	Average[a]	Range[a]
Coke side benchman	18	1.08	0.09— 2.74
Door machine operator	25	2.11	0.04— 6.51
Heater	39	1.07	N.D.— 2.98
Larry car operator	39	3.05	0.28— 8.78
Lidman	61	3.22	0.42—17.89
Luterman	18	2.57	0.25— 4.82
Miscellaneous	18	0.93	0.18— 2.76
Pusher machine operator	23	0.39	N.D.— 0.93
Pusher side benchman	28	2.03	N.D.—14.59
Quench car operator	23	0.94	N.D.—7.20
Tar chaser	27	3.14	0.04—14.70
Total	319	2.08	N.D.—17.89

Note: N.D. — not detected.

[a] mg/m^3

From Fannick, N., Gonshor, L. T., and Shockley, J., Jr., *Am. Ind. Hyg. Assoc. J.*, 33, 461, 1972. With permission.

of PAH. Consequently, no good correlation can be expected between the levels of BaP and those of most other PAH present in the working atmosphere of different industries.

2.   BaP levels can only account for a minor part of the carcinogenic and mutagenic activity of organic extracts from suspended particulate matter.

There are surprisingly few studies involving detailed analysis of PAH in workplace atmospheres. The detailed analysis provides superior information on the chemical environment. It may be used for the characterization of different processes and industries, and may provide a more objective basis for regulatory standards.

II. GAS WORKS AND COKE OVENS

In gas and coke works, large amounts of PAH are released from the ovens in which coal is pyrolyzed. Major exposures to workers occur during charging operations, or result from leakage around the lids or pipes at the top of the ovens or from the oven doors due to incomplete sealing. In most studies of PAH in coke oven atmospheres, BSF has been measured, but data on BaP as well as other PAH are also available.

In an extensive study, Fannick et al.[6] measured BSF at ten by-product coking industries in Pennsylvania. A total of 319 personal samples were taken from 11 different categories of workers, resulting in an overall average exposure of 2.08 mg benzene solubles per cubic meter. As seen in Table 1, the most exposed groups were lidmen, tar chasers, and larry car operators. Individual peak values up to 18 mg/m^3 were found. More than 80% of all samples were above the TLV of 0.20 mg/m^3.

Jackson et al.[7] determined the weight of the cyclohexane-soluble part of the particulates in a coke work. At the battery top, this fraction was 0.4 to 1.7 mg/m^3. At two of the on-battery sampling stations, the BaP content of the cyclohexane-soluble fraction was determined to be 0.78 and 0.63%, respectively.

Table 2
BaP CONCENTRATIONS IN GAS AND
COKE WORKS ($\mu g/m^3$)

Occupation	Mean	Range	Ref.
Gas Works			
General	7.3		9
	2.0		9
Top side work	4.3	0.007—33	8
25 m from oven	0.15	0—1.7	8
General	1.4—4.8		10
Coke Plants			
Top side work	13.5	7—20	8
	9.4	2.9—24.1	8
	5—7	4—18	14
	6.5	1—10	7
	63		15
	23	2—40	16
	22.3		17
	33		17
	55	24—135	18
	37	14—69	18
Various	1.5—3.1		11
	3.8	0.005—7.4	12
	2—32	1—52	13
Larry car operation	9	2.6—22	8
	22	12—43	18
	4.5		17
Push car operation	3	1—44	18
Cleaning	13	6—20	18

An investigation of Swedish gas works in 1964[8] showed an average BSF value of 0.69 mg/m^3 for a total of 134 samples from the top of the ovens. The peak value was 7.6 mg/m^3 and the mean content of BaP was 0.62%.

Many reports on BaP concentrations in coke ovens and gas works have been published. Generally, the mean value is below 10 $\mu g/m^3$, as is shown in Table 2. However, in some cases, higher BaP concentrations have been reported. In Czechoslovakian coke works, Masek[13] found average values of BaP up to 32 $\mu g/m^3$ in close vicinity to the ovens. Thielen[16] reported high BaP concentrations on and under the larry car. The median value of 28 samples was 25 $\mu g/m^3$ and the range was 8 to 78 $\mu g/m^3$. Stationary sampling at the battery top of a Norwegian coke plant showed mean BaP concentrations of 37 and 55 $\mu g/m^3$ in two series, respectively.[18] In this study, the exposure of some workers was also investigated by personal air sampling. The highest BaP exposure was found for the larry car operator (12 to 43 $\mu g/m^3$), the wharfman (21 $\mu g/m^3$), and the jamb cleaner (6 to 18 $\mu g/m^3$). Blome[17] reported BaP values of 22 and 33 $\mu g/m^3$ for lidmen, and 4.5 $\mu g/m^3$ for the larry car operator. These results demonstrate a generally high exposure level for topside work at coke ovens. However, the actual exposure is expected to vary greatly, depending on wind and other climatic conditions, protection of workers, stage of the carbonization process, and other factors.

A detailed study, using CGC has been published for a Norwegian coke plant.[18] Both particulate and gaseous PAH were determined. The volatile components were trapped in ethanol wash bottles cooled with dry ice. Table 3 gives the distribution between gaseous and particulate PAH for ten stationary samples from the battery top of the coke plant.

Table 3
PAH IN THE ATMOSPHERE OF A COKE PLANT; STATIONARY
SAMPLING AT THE BATTERY TOP, FALL 1976 (μg/m^3)

PAH	Sampling site			
	Battery top[a]		Battery top[b]	
	Particulates	Gaseous	Particulates	Gaseous
Naphthalene	—	294.46	—	459.91
2-Methylnaphthalene	—	41.03	—	126.34
1-Methylnaphthalene	—	20.78	—	68.35
Biphenyl	—	7.62	—	17.93
Acenaphtylene	—	47.55	—	113.14
Acenaphtene	—	10.86	—	25.19
Dibenzofuran	—	17.33	—	35.24
Fluorene	—	24.92	—	37.14
Dibenzothiophene	0.77	5.35	4.07	8.43
Phenanthrene	11.24	71.39	59.40	71.82
Anthracene	1.76	16.85	25.83	18.66
Carbazole	1.89	—	4.22	0.46
2-Methylanthracene	0.76	—	9.12	—
1-Methylphenanthrene	1.15	1.00	7.54	1.02
Fluoranthene	32.38	7.60	71.29	1.62
Dihydrobenzo(a and b)fluorene	3.25	—	16.30	—
Pyrene	24.87	3.98	51.56	0.77
Benzo(a)fluorene	6.40	—	21.10	—
Benzo(b)fluorene	0.92	—	28.27	—
4-Methylpyrene	0.68	—	—	—
1-Methylpyrene	1.27	—	5.81	—
Benzo(c)phenanthrene	5.47	—	5.21	0.38
Benzo(a)anthracene	11.32	—	31.09	—
Chrysene/triphenylene	14.49	—	35.55	—
Benzo(b,j, and k)fluoranthene	4.70	—	7.20	—
BeP	3.88	—	6.41	—
BaP	8.02	—	14.51	—
Perylene	1.80	—	3.15	—
o-Phenylenepyrene	3.83	—	3.14	—
Benzo(ghi)perylene	4.20	—	7.53	—
Anthanthrene	1.70	—	1.48	—
Total PAH	146.75	569.72	419.78	986.40

[a]   Particle concentration 5.76 mg/m^3.
[b]   Particle concentration 7.21 mg/m^3.

From Bjørseth, A., Bjørseth, O., and Fjeldstad, P. E., *Scand. J. Work Environ. Health*, 4, 224, 1978. With permission.

Usually, components up to pyrene passed the filter and were identified in the absorption liquid. In extreme cases, it was observed that up to 5% of BaP also passed the filter. For the most part, the concentration of gaseous PAH was found to be of the same order as the particulate PAH, with the concentration ratio varying from 1:1 to 1:3.

On the battery top, the sum of identified PAH in the air was, on the average, 2.8 and 2.7 μg/m^3 during the sampling periods in spring and fall, respectively. The gaseous fraction was dominated by naphthalene and the methylnaphthalenes (47 to 70%). Main components in the particulate fraction, in decreasing order of occurrence, were phenanthrene fluorene, and anthracene. The mean content of BaP in the PAH fraction of the particulate matter was 3.2 and 3.7% during the two sampling periods.

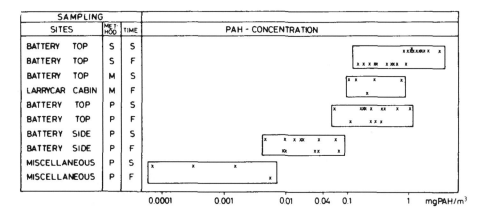

FIGURE 1. Concentration of PAH in particulate matter at a coke plant measured by stationary (S), mobile (M), and personal sampling (P) during spring (S) and fall (F). (From Bjørseth, A., Bjørseth, O., and Fjeldstad, P. E., *Scand. J. Work Environ. Health*, 4, 224, 1978. With permission.)

Table 4
DETERMINATION OF THE BSF IN ALUMINUM PLANTS

Type	Occupation	BSF (mg/m³)		Number of samples	Ref.
		Range	Time weighted average		
Vertical pin	Potman	n.d.— 6.8	2.2	10	19
	Tapper	1.0—63.4	12.0	5	
	Tapping craneman	n.d.— 7.1	3.0	3	
	Pin setter	2.7— 7.5	2.5	2	
	Head pin setter	11.3—19.0	7.5	3	
Horizontal pin	Potman	4.6— 8.7	4.5	4	
	Pindriver	8.3—26.9	12.0	4	
	Tapper	3.0— 3.6	2.0	2	
	Pin puller	2.2— 3.3	1.5	2	
	Anode craneman	35.3—60.	17.5	2	
Prebaked	Potman	n.d.— 0.2	0.2	3	
	Tapper/anode setter	n.d.— 2.1	0.5	2	
	Craneman	0.4— 0.7	0.4	2	
Carbon plant	Operator	n.d.— 0.4	0.2	2	
Prebaked	Craneman	0.7— 1.0	0.8	2	
Anode prod.	Curing fumal operater	0.4— 2.3	1.4	2	
Vertical pin	Stationary sampling	0.2— 1.6	0.42	48, time = 13 min (1968)	8
		0.05—0.34	0.14	36, time = 13 min (1972)	

The concentration of total PAH in particulate matter measured by stationary, mobile, and personal sampling in both spring and fall periods are summarized in Figure 1. Personal sampling gave somewhat lower values than stationary sampling because it also operated during rest periods. As revealed by the figure, PAH exposure varies greatly with job type. The PAH concentration is highest at the battery top and some work operations at this site may result in very high exposure to PAH.

The concentration of PAH on different particle-size fractions has also been studied[18] (See Table 4, Chapter 3). PAH could be detected in all size fractions; however, the bulk of PAH was adsorbed to particles in the range from 7 to 0.9 μm. If particles smaller than 7 μm are classified as respirable, the table shows that 98% of the PAH are in the respirable fraction.

III. ALUMINUM PLANTS

Basically, there are two different technologies for producing aluminum: the Söderberg anode and the prebaked anode technologies. The former, which is used most commonly, is the main cause of PAH exposure in aluminum plants. At the high temperatures used, large amounts of PAH are liberated from the electrodes containing coal tar pitch. These PAH may escape into the work environment. The simplest way of monitoring the PAH exposure, by determination of the BSF, has been applied to a large number of samples from aluminum plants. Table 4 shows some of these data. As revealed by this table, the exposure varies considerably between different occupations and different plants. The values should be compared with the American TLV of 0.2 mg/m³. It is evident that this TLV is heavily exceeded for many workers in the aluminum plants. However, as mentioned previously, the gravimetric benzene-soluble technique shows unsatisfactory performance for measuring exposure to particulate PAH (see Chapter 6, Section II).

The BSF technique measures all soluble material collected on a filter as a single value. Very often workplace atmospheres in aluminum plants contain aliphatic lubricating oils, vegetable oils, or other contaminants which are benzene-soluble and whose presence may be unsuspected. These extraneous benzene-soluble substances, which are relatively innocuous compared to PAH, would tend to overstate the hazard potential arising from BSF measurements.

Several data have been reported on the occurrence of BaP in the work atmosphere of aluminum plants. Some measurements for Söderberg and prebaked aluminum plants are shown in Table 5. The table reveals that there is considerable variation among the different occupations in an aluminum plant. Pin pulling results in the highest exposure of the categories measured. It is also evident from the table that the level of BaP is considerably lower in a prebaked plant than in a Söderberg plant.

Only a few reports have been published where a detailed analysis of individual PAH compound has been made. Bjørseth and Eklund[24] have used a glass CGC/MS computer system for the identification of PAH in an air particulate sample from an aluminum plant. More than 100 PAH and heterocyclic polynuclear aromatic compounds were identified in the sample. The majority of the compounds identified were parent PAH or their methyl, dimethyl, and/or ethyl derivatives. About 30 of these, including all of the main compounds, have been previously reported from CGC.[20] Several of the compounds identified show carcinogenic properties in laboratory animals. These include benz(a)anthracene, benzo(b)- and benzo(k)fluoranthene, BaP, and dibenzanthracenes. PAH are found, however, not only on particulate matter in the aluminum plant. Bjørseth et al.[5,20] observed that the major part of the PAH fraction by weight was always the volatile component (two- and three-ring PAH), such as naphthalene, biphenyl, fluorene, phenanthrene, and their methyl derivatives. Table 6 shows a distribution of PAH between gaseous and particulate phases in a vertical pin Söderberg aluminum plant. Compounds up to pyrene were observed both in the gaseous and the particulate phases.

Similarly, using personal sampling, Andersson et al.[25] observed that a large fraction of PAH passed the filter and could be found in the following adsorbent section (see Table 1, Chapter 5). Even for the benzofluoranthenes about 3% had passed the filter. They explained this finding by the rather high ambient temperature in the vicinity of the reduction pots.

Bjørseth et al.[20] determined the total amount of particulate PAH in different plants for comparison. Figure 2 shows the results for both stationary and personal sampling for a plant with vertical pin Söderberg cells and with closed prebaked anode cells, as well as for an anode plant and the pitch-bin of the anode paste plant. As revealed by the figure, the atmospheric PAH content was significantly lower in the prebaked plant, being in the region of 1 to 10% of the amount in the Söderberg plant. The anode plants and the Söderberg

Table 5

DETERMINATION OF BENZO(A)PYRENE IN ALUMINUM PLANTS

| | Plant type | | | | | |
| | Söderberg | | Prebake | | Sampling | |
Occupation	Range	Mean	Range	Mean	methods	Ref.
Crust break		12			n.a.	8
	0.2—3.9	1.2			n.a.	8
Potman	0.1—2.8	2.0			n.a.	8
	2.1—3.6	2.8	—		Stationary	8
			0.3—0.5		n.a.	19
	9—10		n.d.		n.a.	22
	18—29	—			n.a.	22
		36			Stationary	23
		38			Stationary	23
Pin pullers	10—24	—			Personal	8
	54—117	85			n.a.	19
		53			n.a.	19
	4—40	—			Personal and mobile	5
	28—230	—			n.a.	22
	221—975	—	n.d.			22
Pin craneman		29	0.1	0.1	n.a.	19
Tapping craneman		3.1			n.a.	19
Tapper		3.4			n.a.	20
Flex raisers	23—51	—			n.a.	8
Anode service	4—22				n.a.	8
Anode plant	4—5				Stationary	23
Cathode repair	2—17				n.a.	8
Pitch bin		28			Personal and mobile	5
Crane operation	3.6—19	—			n.a.	22
	11—73	—			n.a.	22
	n.d.—800				Stationary	23
	n.d.—30	13			Stationary	23
Various		2				5
	13.7—22.1	—			n.a.	21
		1.9			Stationary	23
		3.7			Stationary	23
General	0.7—9.0	4.3	n.d.—0.05	—	n.a.	20

potrooms showed airborne PAH levels similar to those found on the battery top of the Norwegian coke plant.[18] In the Söderberg potroom, BaP was 3.1 to 6.6% of the total particulate PAH in the stationary samples, and 4.2 to 9.6% of the total in the personal samples.

Figure 2 demonstrates further that there are considerable variations among the personal samples of different job types. Figure 3 summarizes the values observed for different occupations in the Söderberg aluminum plant.

As is frequently observed in Söderberg plants, the pin pullers have the highest exposure, ranging from 50 to 4, 150 μg PAH per cubic meter, with an average of about 900 μg PAH per cubic meter. This is to be compared with the present administrative TLV of 40 μg PAH per cubic meter in Norway, as indicated in the figure. Low exposure, with a mean of 26 μg PAH per cubic meter, was found for foreman and tappers. Other occupations given in Figure 3 had values between those of the highest- and lowest-exposure jobs mentioned.

It is apparent from the study by Bjørseth et al.[20] that nonhomogeneous distribution of PAH in the atmosphere of an aluminum plant potroom may give rise to significant variation in occupational exposure, even for workers doing the same job. In order to study this,

Table 6
TYPICAL DISTRIBUTION OF PARTICULATE AND GASEOUS PAH IN STATIONARY SAMPLES FROM THE POTROOM FROM AN ALUMINUM PLANT

Compound	Particulate PAH (μg/m^3)	Gaseous PAH (μg/m^3)	Total PAH (μg/m^3)
Naphthalene		15.09	15.09
2-Methylnaphthalene		8.94	8.94
1-Methylnaphthalene		5.10	5.10
Biphenyl		1.51	1.51
Acenaphthylene		14.74	14.74
Acenaphthene		4.11	4.11
Dibenzofuran		<0.03	<0.03
Fluorene		9.24	9.24
9-Methylfluorene		2.65	2.65
2-Methylfluorene		0.48	0.48
1-Methylfluorene		1.30	1.30
Dibenzothiopene		2.09	2.09
Phenanthrene	1.49	43.50	44.98
Anthracene	0.10	3.90	4.00
Carbazole	0.96	—	0.96
2-Methylanthracene	0.37	0.20	0.58
1-Methylphenanthrene	0.10	0.60	0.70
9-Methylanthracene	0.11	<0.03	0.11
Fluoranthene	7.58	6.71	14.29
Dihydrobenzo(a + b)fluorene	0.21	0.20	0.41
Pyrene	5.36	3.29	8.65
Benzo(a)fluorene	1.50		1.50
Benzo(b)fluorene	0.39		0.39
4-Methylpyrene			
1-Methylpyrene	0.47		0.47
Benzo(c)phenanthrene	0.31		0.31
Benzo(a)anthracene	2.28		2.28
Chrysene/triphenylene	4.43		4.43
Benzo(b)fluoranthene	4.64		4.64
Benzo(j/k)fluoranthenes			
BeP	1.45		1.45
BaP	0.85		0.85
Perylene	0.26		0.26
Indeno(1,2,3-cd)pyrene	0.61		0.61
Benzo(ghi)perylene	0.63		0.63
Anthanthrene perylene	<0.02		<0.02
1,2,3,4-Dibenzpyrene	<0.02		<0.02
Coronene	<0.02		<0.02
Total	34.1	123.7	164.9

Adapted from Bjørseth, A., and Bjørseth, O., and Fjeldstad, P. E., *Scand. J. Work Environ. Health*, 4, 212, 1978.

personal sampling was carried out for the same workers on several consecutive days.[5] As is seen from the data of Table 7, there are large variations in the exposure from day to day for the same person doing the same job, occasionally with as high a factor as 75. In some jobs, such as gas cleaning, there are occasionally days with very high exposure. On these occasions, however, workers may use protective aids, thus reducing the real occupational exposure.

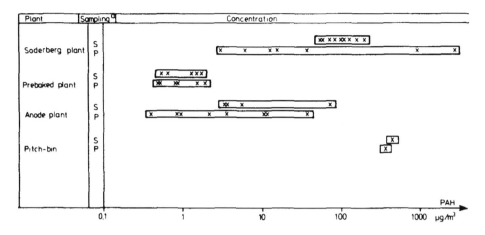

FIGURE 2. Particulate PAH at different sites of an aluminum plant collected by personal (P) and stationary (S) sampling. (From Bjørseth, A., Bjørseth, O., and Fjeldstad, P. E., *Scand. J. Work Environ. Health*, 4, 212, 1978. With permission.)

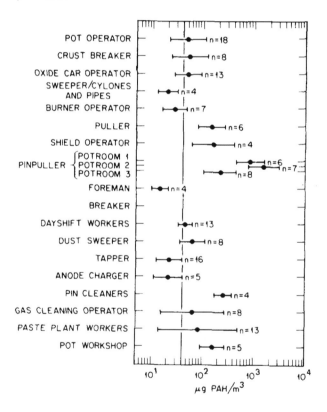

FIGURE 3. Total PAH in an aluminum plant according to job type. (From Gammage, R. B., *Handbook of Polycyclic Aromatic Hydrocarbons*, Bjørseth, A., Ed., Marcel Dekker, New York, 1983, 653. With permission.)

IV. IRON AND STEEL WORKS

In addition to coke production employees, workers in the iron and steel industry are exposed to PAH, originating mainly from coal tar products in contact with molten iron.

In Sweden[8], a series of BaP measurement over more than 10 years have been made at

Table 7
VARIATIONS IN OCCUPATIONAL EXPOSURE IN AN
ALUMINUM PLANT ($\mu g/m^3$)

Job type	Day 1	Day 2	Day 3	Day 4	Day 5
Gas cleaning operator	39.0	12.0	79.9	234	
Gas cleaning operator	47.8	11.3	854		
Foreman	17.1	17.0	18.7		
Dayshift worker		93.9	54.9	29.9	42.8
Dayshift worker		51.2	53.6	31.9	47.0
Dayshift worker		54.0	28.3	37.0	61.6
Sweeper		31.9	93.6	28.6	147
Sweeper		86.8	81.5	45.3	64.2
Tapper	12.7	31.6	16.7		
Tapper	11.3	20.1		34.3	
Pot operator	31.9	82.6	65.2		
Burner operator		70.1	55.1	15.3	
Burner operator			26.1	16.2	25.8
Puller			141	71.1	316
Crust breaker	25.8	49.2	13.8		
Oxide car operator	32.5	63.2	39.2		
Oxide car operator	16.6	54.6	21.8		
Sweeper/cyclones and pipes		19.4		12.8	17.3

Adapted from Bjørseth, A., Bjørseth O., and Fjeldstad, P. E., *Scand. J. Work Environ. Health,* 7, 223, 1981.

blast furnaces, electric steel furnaces, open hearth furnaces, and other places in steel works. The results are summarized in Table 8. There is considerable variation in PAH exposure depending on the sampling site and job type. A great variation in the BaP level with time has been observed using short-time stationary sampling. Figure 4 demonstrates the variation in airborne BaP in the vicinity of a blast furnace during an 8-hr workshift. Peak values probably occur during the tapping of iron from the furnaces.

Tanimura[11] found 1.1 to 1.6 μg BaP per cubic meter at a blast furnace, 0.43 to 0.58 $\mu g/$ m^3 at an electric steel furnace, 0.3 $\mu g/m^3$ at an open hearth furnace, and 0.2 to 0.3 $\mu g/m^3$ at a Bessemer converter. The highest level, 4.2 μg BaP per cubic meter, was found in a rolling mill plant in winter. Blome[26,27] reports 0.05 to 0.5 μg BaP per cubic meter for stationary samples, and 0.33 to 2.8 $\mu g/m^3$ for personal samples at blast furnaces.

Rondia[28] studied the PAH exposure of the workers spraying the inside of ingot molds for steel with a tar-containing product. He found extremely high PAH levels, between 90 and 126 $\mu g/m^3$. In recent years, the use of coal tar pitch in steel plants has decreased in order to reduce these extreme PAH exposures.

PAH have also been reported in a Norwegian iron works.[29] where pig iron is produced in electric furnaces. Total PAH exposure was found to be 17 to 93 $\mu g/m^3$ for the workers tapping the ovens. All other workers were exposed to concentrations below 40 $\mu g/m^3$. On the average, BaP made up 7% of all the PAH components of the particulate fraction. The composition of the PAH fraction in the samples from these iron works was quite similar to that found in the Norwegian coke works.

V. FOUNDRIES

The PAH exposure problems in foundries are principally of the same type as in the iron and steel industry. Different organic materials are used as additives in the molding sand. Starch, oils, and resins are used as binders to improve the compactibility of the sand, and

Table 8
BaP IN SOME SWEDISH IRON AND STEEL WORKS

Sampling site, year	Sampling time	Mean BaP (μg/m³)
Two blast furnaces, 1969	~9 min	0.03—6.2
		(means of 5 series, N = 224)
Two blast furnaces, 1976	1.5—4.2 hr	0.88—1.7
Electric steel works, 1969	~9 min	0.44—2.4
		(means of 3 series, N = 222)
Five electric steel works, 1976—1977		
Smelters[a]	2.5—6 hr	0.02—0.52
Repairing of furnace walls[a]	2—6 hr	0.1—4.9
Mold operators[a]	6—7 hr	2.6
Open hearth furnace, 1977		
Smelters[a]	6—7 hr	0.17
Crane operators[a]	6—7 hr	0.58
Mold operatores[a]	6—7 hr	2.5

[a]   Personal air sampling.

From Lindstedt, G. and Sollenberg, J., *Scand. J. Work Environ. Health*, 8, 1, 1982. With permission.

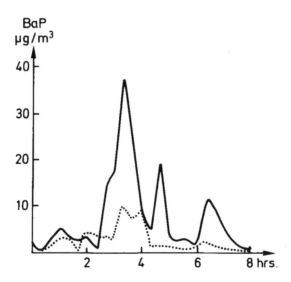

FIGURE 4.   Airborne BaP concentrations at a blast furnace as a function of time over an 8-hr shift at about 15 m (——) and 7 m (...) above ground level.[8]

coal powder, pitch, and wood flour are used as formers of lustrous carbon to avoid iron-sand reactions. PAH may be produced and released during the casting process under the high temperatures involved.

Gwin et al.[30] studied the volatile components generated from green sand, containing 6% bentonite clay, 1% cereal binder, and 3% seacoal. More than 200 PAH components, including BaP were identified but not quantified in the pyrolysis products from a laboratory iron casting experiment. A large number of PAH compounds in gases from iron casting, ranging from naphthalene to benzo(ghi)perylene, have also been identified by Schimberg et al.[31]

Airborne BaP concentration in six Finnish foundries have been investigated by Schimberg, et al.[32]. By grouping the data according to different types of additives, two main distributions were found. Foundries using coal tar pitch had BaP concentrations in the work atmosphere

Table 9
CONCENTRATIONS OF BaP IN IRON FOUNDRY WORK
ATMOSPHERES (μg/m³)

| | Two foundries using coal tar pitch | | Four foundries using coal powder | |
	Mean	Range	Mean	Range
Melting	0.2	<0.01—0.83	0.01	<0.01—0.02
Molding	2.2	0.7—3.3	0.04	<0.01—0.07
Casting	2.1	0.3—5.1	0.09	0.01—0.36
Shake-out	12.6	0.7—5.8	0.10	0.01—0.82
Fettling	0.5	0.15—0.8	0.11	<0.01—0.26
Transport and others	1.4	0.6—3.3	0.04	0.01—0.09

Adapted from Schimberg, R. W., Pfäffli, P., and Tossavainen, A., *Toxicol. Environ. Health*, 6, 1187, 1980.

about 50 times higher than those in the foundries using coal powder as an additive. Table 9 shows the BaP concentration at different working sites of the two groups of foundries studied.[33] The average BaP concentration in foundries using coal tar pitch was 4.9 μg/m³ compared to 0.08 μg/m³ for foundries using coal powder. The large ranges indicate that the hygienic situation in the foundries may vary greatly during a work shift and from day to day. BaP concentrations were especially high during shake-out.

Zdrazil and Picha[34] determined the BaP content in air in two iron foundries. In one of these, the BaP concentrations were 0.12 μg/m³ or less at all working operations. In the other foundry, the highest value observed was 0.47 μg/m³ on tumbling. Higher figures up to 4.1 μg BaP per cubic meter, were found at pressure die castings of aluminum alloys.

Results of BaP measurement in Swedish iron and steel foundries have been reported.[8] Of the 9 samples from an iron foundry and 18 samples from a steel foundry, all but one sample showed values less than 0.06 μg BaP per cubic meter. The maximum value was 0.12 μg BaP per cubic meter.

Recently, Verma et al.[35] determined the PAH content in the air of ten Canadian ferrous and nonferrous foundries, using both stationary and personal sampling. It appeared that high PAH levels were associated with poor general ventilation and the winter season. Certain occupations that have been reported to have an increased risk of lung cancer, such as moulders, casters, and cranemen, were exposed to high amounts of PAH expressed as a percentage of total suspended particles (TSP). In addition, shake-out operators were exposed to high levels of PAH, with a mean BaP concentration of 1.18 μg/m³.

VI. MANUFACTURING OF CARBON ELECTRODES

As has already been mentioned, prebaked or graphite electrodes are used in aluminum smelters, electric steel furnaces, and other metallurgical processes. The raw material from these electrodes is usually petroleum coke with tar or pitch as a binder. During baking, the mixture is heated in ovens to a temperature above 1,000°C, which results in the release of large quantities of PAH from the electrode mass. In the second step, the graphitization, which involves heating up to 2700°C, little PAH is liberated since the PAH already have been driven off during the first heating.

BaP measurements were made in two Swedish graphite electrode plants.[8] Close to the ovens, the median of 27 samples was 5 μg BaP per cubic meter in one plant, but in the other, BaP did not exceed 1.3 μg/m³. Peak values were observed in the pitch impregnation department in both plants with 19 to 40 $\mu g/m^3$. In the graphitization department, only 3 samples out of 24 were above 0.1 $\mu g/m^3$.

Griest et al.[36] investigated PAH exposure during laboratory scale graphite production. Stationary samples showed BaP values of 0.09 to 0.72 $\mu g/m^3$ (mean: 0.26 $\mu g/m^3$), while personal samples varied from 0.28 to 1.35 μg BaP per cubic meter with an average of 1.0 $\mu g/m^3$. Particle size analysis demonstrated that 70% of airborne particulate matter was less than 10 μm in geometric diameter, and hence respirable.

The results of an investigation of PAH exposure in a Norwegian anode plant for making prebaked electrodes are included in Figure 2.[20] The average concentration of total particulate PAH obtained by stationary sampling was 4 $\mu g/m^3$ at four different sites. A maximum value of 73 $\mu g/m^3$ was determined in a sample from the atmosphere above the anode press. The particulate PAH exposure of four workers was measured by personal sampling. There were two jobs with particularly high exposure risk, i.e., coke packing and working in the pitch-bin. The coke packer was exposed to 3.7 μg PAH per cubic meter. In the pitch bin, both stationary and personal sampling gave PAH concentrations in the order of 400 $\mu g/m^3$ in the particulate matter (see Figure 2).

In a Norwegian Söderberg paste plant,[5] two stationary samples showed the extremely high concentrations of total particulate PAH of 1560 and 3900 $\mu g/m^3$. The anode paste plant was, however, highly automated, and occupational exposure was mainly connected to control and occasional repair. The personal exposure may therefore vary within wide ranges. Values from 5 to 1200 $\mu g/m^3$, with a geometric mean of 87 $\mu g/m^3$, were shown for 12 personal samples, taken over approximately 6 hr.

VII. ROOFING, PAVING, AND INSULATING

Asphalt and coal tar pitch are widely used in roofing and paving operations, and to some extent in insulating coatings. The concentrations of PAH in coal tar pitch is usually two to three orders of magnitude higher than those found in petroleum-derived asphalt.[37,38] Malai-yandi et al.[38] determined the content of 11 PAH, including BaP, in five roofing and paving asphalts and three coal tar pitch samples of the grade commonly used in roofing. The sum of PAH was 0.17 to 0.83 mg/g in the asphalt samples and 265 to 500 mg/g in the coal tar pitches. Since these products are handled at elevated temperatures, workers may be exposed to airborne PAH evaporating from the materials.

Hammond et al.[39] determined the exposure of roofers and waterproofers to BaP in pitch and asphalt fumes. The men working in various jobs wore masks during the entire work shift. The filters in the masks were subsequently analyzed for BaP. A summary of the findings is given in Table 10. As is shown in the table, the amount of BaP collected in the masks varied with type of job. In most cases, however, considerable inhalation of BaP may take place.

In a Polish study, airborne PAH was determined in an asphalt production plant.[40] During the oxidation of asphalt by airblowing particulate benzene-soluble matter reached levels of up to 17 mg/m³ when hot asphalt was poured. The BaP concentration was never above 0.05 $\mu g/m^3$, except when asphalt was burned off tubes, where a BaP concentration of 0.7 $\mu g/m^3$ was measured. BeP and perylene had somewhat higher concentrations than BaP, which is the normal situation for asphalt products. Von Lehmden et al.[41] found high amounts of pyrene and anthracene in the exhaust gases from asphalt air-blowing, but were unable to detect any larger PAH.

During the asphalt paving of German highways, Zorn[42] found very high BaP exposure for some workers (50 to 350 $\mu g/m^3$) when an asphalt-tar mixture was used. Blome,[26] however, did not find BaP concentrations above 0.05 $\mu g/m^3$ during the handling of asphalt in paving

Table 10
BaP INHALED IN VARIOUS
ROOFING OCCUPATIONS

	μg BaP in masks per 7-hr working days	
	Range[a]	Average
Kettleman	6.0—83.3	31.0
Hose tender	10.5—46.2	28.3
Hot carrier	n.d.—48.7	24.0
Mop man	0.9—4.7	2.9
Felt layer	n.d.—2.5	1.4
Shovelman	22.9—70.7	53.0
Sprayman	25.2	25.2
Scraper	4.2—135.0	51.8
All studies	n.d.—135.0	16.7

[a]   n.d. is not detected.

From Hammond, E. C., Selikoff, I. J., Lawther, P. L., and Seidman, H., *Ann. N.Y. Acad. Sci.*, 271, 116, 1976. With permission.

and insulating operations. Similarly, in a Norwegian study,[43] the total PAH exposure for asphalt-paving workers was as low as 1.3 to 27.2 μg/m³.

Bonnet[44] studied a process in which tar was melted and mixed with cork powder to be spread out on walls as insulating material. BaP concentrations higher than 1 mg/m³ were observed. To the best of our knowledge, this surpasses all other BaP measurements ever made in an occupational environment. High values have also been reported by Sawicki et al.,[45] who observed up to 78 μg BaP per cubic meter at the tarring of a sidewalk and 14 μg BaP per cubic meter at roof tarring. These operations also involve the handling of hot tar close to the worker.

Larson[46] studied exposure at eight plants where metallic pipes were coated by hot coal tar and wrapped with a special paper to provide an anticorrosive coating. Six different categories of workers were distinguished, all of which were exposed to more than 0.2 mg/m³ BSF. The most exposed category, the coating operators, had a mean exposure of 6.53 mg/m³, the highest individual exposure being 23.7 mg/m³. The mean exposure of all personnel at the pipeline-coating operations was found to be 1.89 mg/m³, almost equal to that in the coke oven industry.[6] In a laboratory study using the same tar mixture under identical conditions, 80% of the emissions were found to consist of particles smaller than 10 μm in size. The emissions were dominated by the lower PAH. BaP accounted for 0.46% of the benzene solubles in the particulates.

A thorough investigation of the exposure of roofing and paving crew members by PAH-laden fumes derived from coal tar pitch and asphalt has been reported by Malaiyandi et al.[38] Personal samples were taken using a sampling system consisting of a glass fiber filter followed by a Tenax-GC adsorbent packing in order to trap gaseous PAH and those PAH blown off the filter.

A total of 11 PAH, ranging from fluoranthene to indeno(1,2,3-cd)pyrene, were analyzed by HPLC with fluorescence detection. Table 11 shows the concentrations of BaP and the sum of identified four- and six-ring PAH for different occupations. During asphalt paving and roofing, the average "total" PAH concentration was 23 μg/m³. One filter cartridge from a kettleman at an asphalt roofing site had an unusually high PAH level compared to the others. At the coal tar pitch roofing sites, the average PAH concentration was 25 times

Table 11

PAH EXPOSURE DURING PAVING AND ROOFING OPERATIONS

Site description	Occupation	Sample volume (ℓ)	BaP ($\mu g/m^3$)	Total PAH ($\mu g/m^3$)
Asphalt paving 1	Machine driver	580	0.02	10.2
	Level wheel operator	580	0.01	5.6
Asphalt paving 2	Machine driver	870	Trace	13.0
	Level wheel operator	870	0.01	4.3
Asphalt roofing 1	Applicator	470	0.08	17.6
	Kettleman	540	0.43	112.5
Asphalt roofing 2	Applicator	820	0.04	17.3
	Kettleman	840	0.04	14.5
Asphalt roofing 3	Applicator	620	0.05	16.4
	Kettleman	680	0.08	20.4
Coal tar pitch	Applicator	590	0.04	228.1
Roofing 1	Kettleman	780	4.22	1325.4
Coal tar pitch	Applicator	510	1.22	547.7
Roofing 2	Kettleman	540	0.62	337.1
Coal tar pitch	Applicator	340	0.93	237.4
Roofing 3	Kettleman	440	11.30	896.8

Adapted from Malaiyandi, M., Benedek, A., Holko, A. P., and Bancsi, J. J., *Polynuclear Aromatic Hydrocarbons: Physical and Biological Chemistry,* Cooke, M., Dennis, A. J., and Fisher, G. L., Eds., Battelle Press, Columbus, Ohio, 1982, 471.

higher than at the asphalt roofing sites. Kettlemen at two of the sites were exposed to significantly higher PAH concentrations than the applicators. From the separate analyses of the filters and Tenax-GC packing, it appears that in the case of asphalt roofing activity, most of the PAH are in the vapor phase, whereas the PAH emanating from coal tar pitch are in the form of condensed particulate matter or aerosols.

VIII. COAL CONVERSION PLANTS

A. Coal Gasification Plants

Little data are available concerning the types and quantities of PAH in the atmosphere at coal gasification plants. Only few commercial coal gasifiers exist today but there are a number of experimental and pilot units. Nevertheless, some useful comparisons can be made between the nature of process stream tars produced by gasifiers and airborne emissions from coking operations.

As shown in Figure 5, the tar from the electrostatic precipitator of a low-Btu coal gasifier (100 to 200 Btu/ft^3) is quite similar in composition to the PAH profile for tarry fumes associated with coke oven emissions.[47] This is not surprising, as the gasification process employs reaction conditions and temperatures similar to those used in the coking process.

Recently, pilot-scale gasifiers representative of each of the three generic types of gasification were studied.[48] Area and personal sampling were performed for PAH using silver membrane filters and a Chromosorb back-up,. Material eluted from the samplers was analyzed for 27 PAH by CGC, GC/MS, or HPLC.

Total vapor and aerosol concentrations measured in various operation areas for each site are presented in Table 12. Naphthalene and its methylderivatives were found to dominate. PAH with more than three rings were detected only in trace amounts or not at all. PAH concentrations were much greater for the fixed-bed gasifier than for the other two types, and the levels were somewhat higher for the fluidized-bed gasifier than for the entrained-bed gasifier.

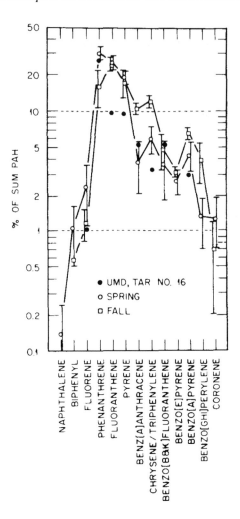

FIGURE 5.   PAH profiles at a coke plant in spring
and fall sampling and of a low Btu coal gasifier tar.
(From Gammage, R. B., *Handbook of Polycyclic
Aromatic Hydrocarbons*, Bjørseth, A., ed., Marcel
Dekker, New York, 1983, 653. With permission.)

Personal samples from various categories of workers showed that exposure associated
with fixed-bed gasification was greater than that associated with the other two processes
(Table 13). The only exception to this was the exposure of a maintenance worker in the
fluidized-bed gasification plant. Most exposures were to two- or three-ring PAH.

Wipe samples indicated that workers might be exposed to PAH of higher molecular weight,
and hence potentially carcinogenic PAH, through contact with the skin.

B. Coal Liquefaction Plants

Comprehensive surveys at three pilot liquefaction facilities have recently been conducted
in the U.S.[47] Both stationary and personal samples were collected using a sampling device
consisting of a filter and a Chromosorb adsorbent back-up. Table 14 shows the results for
stationary samples from different plant locations. Two- and three-ring PAH are dominant,
comprising 45 to 92% of the total polycyclic aromatic compounds detected. The concentration
of five- and six-ring PAH was below the detection limit. It is also noteworthy that quinoline,
a well-recognized carcinogen, is abundant in some samples. Since quite different products

Table 12

**TOTAL CONCENTRATIONS OF PAH FOR AREA SAMPLES OF
COAL-GASIFICATION PLANTS**

Plant	Area	Number of samples	Geometric mean concentration ($\mu g/m^3$)
Entrained-bed	Gasifier combustor	4	0.4
	Gasifier reductor	1	0.4
	Gas duct	2	<0.1
	Induced-"draft" fan	2	0.3
	Sludge thickener	1	<0.1
	Cyclone scrubber	1	0.2
	Control room	2	0.2
	Total	13	0.2
Fixed-bed	Top of gasifier	3	22.7
	Poke hole	3	17.3
	Detarrer/deoiler	4	18.1
	Oil/liquor separator	1	74.2
	Ask pan	1	42.1
	Day tar/oil tank	2	20.6
	Tar pump	2	72.4
	Gas compressor	1	63.6
	Control room	2	20.5
	Total	19	26.8
Fluidized-bed	Gasifier-5th level	3	0.6
	Gasifier-6th level	4	5.4
	Gas compressor	4	1.7
	Filter/stainer	2	0.9
	Control room	4	2.9
	Total	17	2.0

Adapted from Cubit, D. A. and Tanika, R. K., Industrial Hygiene Assessment of Coal Gasification Plants, Final Summary Rep. by Enviro Control Division, Dynamac Corporation, for the National Institute of Occupational Safety and Health, March, 1982.

are being produced or processed in different areas of a coal liquefaction plant, it is not surprising that the samples from different plant areas show substantial variability in the relative proportion of the PAH.

The results presented in Table 15 are for personal sampling in the coal-preparation area. Even in this single area, a great variation in the relative proportions of PAH as well as the quantitative exposures to total PAH is observed. However, in all three samples naphthalene and its methyl derivatives comprise the major constituent PAH. In contrast, wipe samples from contaminated equipment and tools contained larger amounts of four- and higher-ring PAH. Therefore, dermal rather than respiratory exposure could be considered to be a greater health concern.

IX. CHIMNEY SWEEPING

As pointed out in Chapter 1, chimney sweeping was the first occupation with PAH exposure in which an increased risk of cancer was observed. Surprisingly, there is little data available on the PAH exposure of chimney sweeps. A recent paper by Bagchi and Zimmerman[49] reports an exposure during cleaning from the top of the chimney of 14.5 mg benzene solubles per m^3 and 40 μg BaP per cubic meter in the breathing zone of the chimney sweep. Cleaning from the hearth resulted in considerably less exposure. Cadez[50] reported CTPV concentrations 30 to 96 times in excess of the TLV (0.2 mg/m^3) for chimney sweeps in Slovenia.

Table 13
TOTAL CONCENTRATIONS OF PAH IN PERSONAL SAMPLES AT COAL-GASIFICATION PLANTS

Plant	Personnel	Number of samples	Geometric mean concentration ($\mu g/m^3$)
Entrained-bed	Equipment operators	8	2.6
	Maintenance personnel:		
	Welder/boilermaker	4	6.8
	Pipefitter	2	6.5
	Laborer	2	0.4
	Chemical technician	2	0.4
	Instrument technician	1	1.9
Fixed-bed	Operators	4	19.3
	Maintenance personnel:		
	Millwright	2	58.0
	Utility helper	2	15.3
	Maintenance	2	88.5
	Electrician	1	389.4
	Foreman	1	35.1
Fluidized-bed	Lower-level technician	7	4.2
	Upper-level technician	5	5.3
	Maintenance personnel	1	60.4

Adapted from Cubit, D. A. and Tanika, R. K., Industrial Hygiene Assessment of Coal Gassification Plants, *Final Summary Rep.* by Enviro Control Division, Dynamac Corporation, for the National Institute of Occupational Safety and Health, March 1982.

Table 14
INDIVIDUAL PAH BY PERCENTAGE IN AREA AIR SAMPLES TAKEN AT THE SOLVENT REFINED PILOT PLANT (SRC-II PROCESS), FORT LEWIS, WASHINGTON

Compound	Coal preparation area			Mineral separation	Dissolver preheater	Solvent recovery
Naphthalene	10.1	1.1	10.3	4.0	28.3	11.3
Quinoline	24.3	2.2	1.0	6.5	34.2	43.1
2-Methylnaphthalene	14.4	2.6	9.1	15.6	10.7	11.3
1-Methylnaphthalene	1.8	0.7	1.2	0.5	1.6	7.2
Acenaphthalene	1.1	ND	ND	ND	ND	0.1
Acenaphthene	ND	1.5	2.0	2.2	0.3	1.3
Fluorene	7.8	14.6	3.2	3.8	3.2	3.9
Phenanthrene, anthracene	24.3	71.5	52.6	19.3	16.2	17.3
Acridine	3.2	ND	ND	ND	ND	ND
Carbazole	6.7	0.7	9.1	12.9	1.1	0.2
Fluoranthene	3.0	1.8	5.3	10.3	1.1	0.6
Pyrene	2.7	0.7	3.2	11.1	1.1	0.7
Benzo(b)fluorene	0.5	0.4	0.8	4.7	0.4	1.3
Benzo(b)fluorene	ND	2.2	0.8	6.1	ND	1.5
Benz(a)anthracene	0.04	ND	0.8	1.7	1.6	ND
Chrysene, triphenylene	0.02	ND	0.8	0.9	ND	ND
Total PAH ($\mu g/m^3$)	56.3	27.4	50.6	224	18.7	92.9

Note: ND = compound not detected.

From Gammage, R. B., *Handbook of Polycyclic Aromatic Hydrocarbons*, Bjørseth, A., Ed., Marcel Dekker, New York, 1983, 653. With permission.

Table 15
**INDIVIDUAL PAH BY PERCENTAGE IN
PERSONAL AIR SAMPLES TAKEN IN THE
COAL PREPARATION AREA OF THE SOLVENT
REFINED COAL PILOT PLANT (SRC-II)**

Compound	Coal preparation worker		
	Welder	Operator	Operator
Naphthalene	59.9	0.5	18.4
Quinoline	3.9	18.9	ND
2-Methylnaphthalene	10.9	49.5	49.0
1-Methylnaphthalene	1.8	5.4	2.1
Acenaphthalene	3.1	13.7	ND
Acenaphthene	ND	0.4	1.3
Fluorene	7.0	3.2	10.4
Phenanthrene, anthracene	12.4	ND	15.8
Acridine	0.05	0.2	ND
Carbazole	ND	0.5	ND
Fluoranthene	0.7	6.8	ND
Pyrene	1.6	ND	0.5
Benzo(a)fluorene	0.3	0.5	ND
Benzo(b)fluorene	ND	ND	ND
Benz(a)anthracene	ND	ND	0.5
Chrysene, triphenylene	ND	ND	ND
Total PAH ($\mu g/m^3$)	127.0	19.0	38.6

Note: ND = compound not detected.

From Gammage, R. B., *Handbook of Polycyclic Aromatic Hydrocarbons,* Bjørseth, A., Ed., Marcel Dekker, New York, 1983, 653. With permission.

X. EXPOSURE TO EXHAUST FROM COMBUSTION ENGINES

The PAH profiles in the particulate phase of emissions from gasoline and diesel combustion engines have been studied extensively,[51,52] and a great number of PAH have been identified in motor vehicle exhaust. There is, however, little data available on occupational exposure to PAH from combustion engine exhaust.

In Sweden, exposure to BaP, originating mainly from exhaust gases of trucks and other vehicles, has been determined in iron mines since the early 1960s.[8] During the first period, 1960 to 1969, 679 short-time samples were taken during 7 different investigations. Without any traffic near the sampling sites, the mean BaP values never exceeded 0.01 $\mu g/m^3$. When traffic was passing, means between 0.01 and 0.04 $\mu g/m^3$ were observed. The highest 10-min value was 0.55 $\mu g/m^3$.

After 1970, samples were taken continuously on the sampling site for 8 to 32 hr, with the filters being replaced after a few hours to avoid clogging by dust. The results, which are summarized in Table 16, show that exposure to BaP is quite low, and similar to the mean exposure in large urban communities.

In a recent Finnish study,[53] PAH concentrations were determined in a mine atmosphere polluted by exhaust from diesel engines operating underground. Total PAH varied between 0.18 and 0.80 $\mu g/m^3$ with an average of 0.46 $\mu g/m^3$. The relative BaP content was, on the average, 3.4% of the total PAH. This also demonstrates that PAH exposure from exhaust of combustion engines in underground mines is low. Other air pollutants in the mines such

Table 16
BaP ANALYSES (TIME-WEIGHTED AVERAGES) FROM
THREE SWEDISH IRON MINES (1975)

Mine	Number of sampling sites	Number of samples	Sampling time/site (hr)	BaP (µg/m³) Mean	BaP (µg/m³) Range
A	25	180	12—33	0.009	0—0.058
B	25	104	8—32	0.008	0—0.032
C	5	32	25—29	0.005	0.003—0.006

From Lindstedt, G. and Sollenberg, J., *Scand. J. Work Environ. Health*, 8, 1, 1982. With permission.

Table 17
BaP IN SWEDISH AUTOMOBILE REPAIR SHOPS AND GARAGES

Location	Number of sampling sites	Number of samples	Sampling time (hr)	BaP (µg/m³) Mean	BaP (µg/m³) Maximum
Repair shop A	4	24	1	0.028	
Repair shop B	4	24	1	0.044	0.088
Repair shop C	4	24	8	0.035	0.070
Industrial garage	5	60	4	0.004	0.024
Underground public	3	60	2.5	0.011	0.029

From Lindstedt, G. and Sollenberg, J., *Scand. J. Work Environ. Health*, 8, 1, 1982. With permission.

as different dusts, nitrogen oxides, and radon may be greater health hazards, compared to PAH.

In the 1960s, BaP samples were taken in some Swedish tunnels and rock chambers under construction.[8] In five investigations, 164 short-time samples were analyzed. The BaP means were between 0.012 and 0.063 µg/m³. The highest value found for a 10-min sample was 0.11 µg BaP per cubic meter.

During 1964 to 1970, BaP concentrations were determined in some automobile repair shops and garages in Sweden.[8] As Table 17 demonstrates, the BaP levels found in the garages were similar to those found on a city street. In car repair shops, the BaP concentrations were somewhat higher.

Similarly, the concentrations of BaP in London bus garages have not been found to be particularly high, even though they contain much black smoke.[54] In measurements made in the 1950s it was in fact difficult to detect the contributions of buses to the polynuclear aromatic hydrocarbon content of the garage air against the background from coal fire sources.[55] Now, with this background source removed, emissions from the buses, stand out more clearly, but they are still of small magnitude.

Blome[26] studied the exposure of officers at custom checkpoints to PAH from automobile exhaust. The car frequency was 380 to 650 /hr. For stationary samples, the BaP concentration varied between 2 and 15 ng/m³; for personal samples, the BaP concentration was below 10 ng/m³.

XI. SUMMARY AND CONCLUSION

TLVs for PAH in the workplace atmosphere exist in several countries. They are based on the BSF of the particulates, BaP or total PAH in the air. The PAH measurements from all types of workplaces have recently been reviewed and evaluated by Lindstedt and Sollenberg[8] who also ranked the occupations according to the degree of exposure. Since most investigations have been on BaP only, these values served as a basis, but other analyses e.g., BSF or total PAH, would very likely lead to the same results. The authors divided the exposure into several classes, and a slightly modified version is presented below.

1. **Very high exposure (>10 μg BaP per cubic meter)**
   Gas and coke works (topside work)
   Aluminum (Söderberg) plants (certain jobs)
   Manufacturing of carbon electrodes (pitch-bin workers)
   Handling of molten tar and pitch (roofing, paving, insulating, coating, etc.)
   Chimney sweeping (from top)
2. **Relatively high exposure (1 to 10 μg BaP per cubic meter)**
   Gas and coke works in general (side work)
   Blast furnaces
   Steel works (some jobs)
   Manufacturing of carbon electrodes (in general)
   Aluminum plants (in general)
   Roofing and paving using asphalt mixed with tar
3. **Moderate exposure (0.1 to 1 μg BaP per cubic meter)**
   Steel works (in general)
   Foundries (certain jobs)
   Manufacturing of Söderberg electrode paste
4. **Low exposure (0.01 to 0.1 μg BaP per cubic meter)**
   Asphalt production and handling
   Aluminum potrooms with prebaked electrodes
   Foundries (in general)
   Construction of tunnels
5. **Very low exposure (<0.01 μg BaP per cubic meter)**
   Iron mines
   Garages
   Custom checkpoints

Little data are available on the PAH exposure of workers in coal-conversion and shale oil facilities. However, exposure to higher PAH. Such as BaP, in the working atmosphere seems to be low compared to potential exposure through skin contact.

REFERENCES

1. Occupational Safety and Health Administration, U.S. Department of Labor, Exposure to coke oven emissions: occupational safety and health standards, *Fed. Regist.*, 41, 46742, 1976.
2. Threshold Limit Values for Chemical Substances and Physical Agents in the Workroom Environments with Intended Changes for 1983—84, American Conferences of Governmental Industrial Hygienists, Cincinnati, 1983.
3. National Institute for Occupational Safety and Health, NIOSH Criteria for a Recommended Standard: Occupational Exposure to Coal Tar Products, National Institute for Occupational Safety and Health, U.S. Department of Health, Education and Welfare, Cincinnati, 1977.
4. **Shabad, L. M.,** On the so-called MAC (maximum allowable concentrations) for carcinogenic hydrocarbons, *Neoplasma*, 22, 459, 1975.

5. **Bjørseth, A., Bjørseth, O., and Fjeldstad, P. E.,** Polycyclic aromatic hydrocarbons in the work atmosphere. Determination of area-specific concentrations the job-specific exposure in a vertical pin Söderberg aluminum plant, *Scand. J. Work Environ. Health,* 7, 223, 1981.

6. **Fannick, N., Gonshor, L. T., and Shockley, J., Jr.,** *Exposure to coal tar pitch volatiles at coke ovens, Am. Ind. Hyg. Assoc. J.,* 33, 461, 1972.

7. **Jackson, J. O., Warner, P. O., and Mooney, T. F., Jr.,** Profiles of benzo(a)pyrene and coal tar pitch volatiles at and in the immediate vicinity of a coke oven battery, *Am. Ind. Hyg. Assoc. J.,* 35, 276, 1974.

8. **Lindstedt, G. and Sollenberg, J.,** Polycyclic aromatic hydrocarbons in the occupational environment, *Scand. J. Work Environ. Health,* 8, 1, 1982.

9. **Kreyberg, L.,** 3,4-Benzpyrene in industrial air pollution: some reflexions, *Br. J. Cancer,* 13, 618, 1959.

10. **Lawther, P. E., Commins, B. T., and Waller, R. E.,** A study of the concentration of polycyclic aromatic hydrocarbons in gas works retort houses, *Br. J. Ind. Med.,* 22, 13, 1965.

11. **Tanimura, H.,** Benzo(a)pyrene in an iron and steel works, *Arch. Environ. Health,* 17, 172, 1968.

12. **Petrova, N. V.,** 3,4-Benzpyrene pollution of the atmosphere in shops of the Krivoy-Rog by-product coke plant (in Russian), *Gig. Tr. Prof. Zabol.,* 13, 45, 1969; *Chem. Abstr.,* 72, 47108t, 1970.

13. **Masek, V.,** Benzo(a)pyrene in the workplace atmosphere of coal and pitch coking plants, *J. Occup. Med.,* 13, 193, 1971.

14. **British Coke Research Association,** The determination of polynuclear aromatic hydrocarbons in air particulate matter, *Coke Res. Rep.,* 1973, 76.

15. **Szyja, J.,** Untersuchungen über den 3,4-Benzpyrengehalt an Arbeitsplätzen der Pechofenbatterie einer Kokerei und in Körperflüssigkeiten der Arbeiter, *Z. Gesamte Hyg. Ihre Grenzgeb.,* 7, 440, 1977.

16. **Thielen, R. G.,** Benzo(a)pyren-Konzentrationen bei Arbeiten auf Koksbatterien, *Verh. Dtsch. Ges. Arbeitsmed.,* 19, 267, 1979.

17. **Blome, H.,** Messungen polyzyklischer aromatischer Kohlenwasserstoffe an Arbeitsplätzen — Beurteilung der Ergebnisse, *Staub-Reinhalt. Luft,* 41, 225, 1981.

18. **Bjørseth, A., Bjørseth, O., and Fjeldstad, P. E.,** Polycyclic aromatic hydrocarbons in the work atomosphere. II. Determination in a coke plant, *Scand. J. Work Environ. Health,* 4, 224, 1978.

19. **Shuler, P. J. and Bierbaum, P. J.,** Environmental Survey of Aluminum Reduction Plants, Publ. No. 74-101, National Institute for Occupational Safety and Health, U.S. Department of Health, Education and Welfare, Cincinnati, 1974.

20. **Bjørseth, A., Bjørseth, O., and Fjeldstad, P. E.,** Polycyclic aromatic hydrocarbons in the work atmosphere. I. Determination in an aluminum reduction plant, *Scand. J. Work Environ. Health,* 4, 212, 1978.

21. **Filatova, A. S., Kyz'minykh, A. I., Vedernikova, F. D., and Solomennikova, N. S.,** Method for determining 3,4-benzpyrene discharged in sublimation of anodes of electrolysis shops of aluminum plants (in Russian), *Gig. Sanit.,* 31, 55, 1966; *Chem. Abstr.,* 65, 6738h, 1966.

22. **Adamiak-Ziemba, J., Ciosek, A., and Gromiec, J.,** The evaluation of exposure to harmful substances emitted in the process of the production of aluminum in selfbaking anodes (in Polish), *Med. Pr.,* 28, 481, 1977.

23. **Konstantinov, V. G. and Kyz'minykh, A. I.,** Resinous substances and 3,4-benzpyrene in the air of electrolysis shops of aluminum plants and their role in carcinogenesis (in Russian), *Gig. Sanit.,* 36, 39, 1971; *Chem. Abstr.,* 74, 145994q, 1971.

24. **Bjørseth, A. and Eklund, G.,** Analysis for polynuclear aromatic hydrocarbons in working atmospheres by computerized gas chromatography-mass spectrometry, *Anal. Chim. Acta,* 105, 119, 1979.

25. **Andersson, K., Levin, J.-O., and Nilsson, C.-A.,** Sampling and analysis of particulate and gaseous polycyclic aromatic hydrocarbons from coal tar sources in the working environment, *Chemosphere,* 12, 197, 1983.

26. **Blome, H.,** Polyzyklische Aromatische Kohlenwasserstoffe (PAH) am Arbeitsplatz, BIA-Rep, 3/83, Berufsgenossenschaftliches Institut für Arbeitssicherheit, Sankt Augustin, West Germany, 1983.

27. **Blome, H. and Baus, K.,** Konzentrationen polyzyklischer aromatischer Kohlenwasserstoffe (PAH) bei Herstellung und Verwendung von Pyrolyseprodukten aus organischem Material, *Staub-Reinhalt. Luft,* 43, 367, 1983.

28. **Rondia, D.,** La solution rélle d'un problème d'hygiène dans un acièrie, *Arch. Mal. Prof. Med. Trav. Secur. Soc.,* 25, 403, 1964.

29. **Bjørseth, A., Bjørseth, O., and Fjeldstad, P. E.,** Polycyclic Aromatic Hydrocarbons in Work Atmosphere, Tech. Hygienic Rep. No. 3, PAH at Norsk Jernverk A/S (in Norwegian), Institute of Occupational Health, Oslo, 1978.

30. **Gwin, C., Scott, W., and James, R.,** A preliminary investigation of the organic chemical emissions from green sand pyrolysis, *Am. Ind. Hyg. Assoc. J.,* 37, 685, 1976.

31. **Schimberg, R. W., Pfäffli, P., and Tossavainen, A.,** Profilanalyse von polycyclischen aromatischen Kohlenwasserstoffen in Eisengiessereien, *Staub-Reinhalt. Luft,* 38, 273, 1978.

32. **Schimberg, R. W., Pfäffli, P., and Tossavainen, A.,** Polycyclic aromatic hydrocarbons in foundries, *Toxicol. Environ. Health,* 6, 1187, 1980.

33. **Schimberg, R. W.**, Industrial hygienic measurements of polycyclic aromatic hydrocarbons in foundries, in *Chemical Analysis and Biological Fate: Polynuclear Aromatic Hydrocarbons*, Cooke, M. and Dennis, A. J., Eds., Battelle Press, Columbus, Ohio, 755, 1982.

34. **Zdrazil, J. and Picha, F.**, Carcinogenic hydrocarbons, especially 3,4-benzpyrene in foundries, *Slevarenstvi*, 13, 198, 1965. *Chem. Abstr.*, 66, 108032p, 1967.

35. **Verma, D. K., Muir, D. C. F., Cuncliffe, S., Julian, J. A., Vogt, J. H., Rosenfeld, J., and Chovil, A.**, Polycyclic aromatic hydrocarbons in Ontario foundry environments, *Am. Occup. Hyg.*, 25, 17, 1982.

36. **Griest, W. H., Kubota, H., and Eatherly, W. P.**, Characterization of PAH-containing fugitive emissions from a laboratory scale graphite production operation, *Light Met.*, 2, 519, 1978.

37. **Wallcave, L., Garcia, H., Feldman, R., Lijinski, W., and Shubik, P.**, Skin tumorigenesis in mice by petroleum asphalts and coal tar pitches of known polynuclear aromatic content, *Toxicol. Appl. Pharmacol.*, 18, 41, 1971.

38. **Malaiyandi, M., Benedek, A., Holko, A. P., and Bancsi, J. J.**, Measurement of potentially hazardous polynuclear aromatic hydrocarbons from occupational exposure during roofing and paving operations, in *Polynuclear Aromatic Hydrocarbons: Physical and Biological Chemistry*, Cooke, M., Dennis, A. J., and Fisher, G. L., Eds., Battelle Press, Columbus, Ohio, 1982, 471.

39. **Hammond, E. C., Selikoff, I. J., Lawther, P. L., and Seidman, H.**, Inhalation of benzpyrene and cancer in man, *Ann. N.Y. Acad. Sci.*, 271, 116, 1976.

40. **Adamiak-Ziemba, J. and Kesy-Dabrowska, I.**, Pollution of air with benzene soluble substances and polycyclic aromatic hydrocarbons at asphalt production (in Polish), *Med. Pr.*, 23, 617, 1972.

41. **von Lehmden, D. J., Hangebrauck, R. P., and Meeker, J.**, Polynuclear hydrocarbon emissions from selected industrial processes, *J. Air Pollut. Control Assoc.*, 15, 306, 1965.

42. **Zorn, H.**, Berufliche und Umweltgefährdung durch polyzyklische aromatische Kohlenwasserstoffe (PAH) in Erdöl- und Kohledestillaten bei ihrer technologischen Anwendung in der Industrie und im Strassenbau, *Arbeitsmed. Sozialmed. Präventivmed.*, 13, 619, 1978.

43. **Bjørseth, A., Bjørseth, O., and Fjeldstad, P. E.**, Polycyclic Aromatic Hydrocarbons in Work Atmosphere, Tech. Hygienic Rep. No. 4, PAH Exposure at Production and Paving of Asphalt and Oil Gravel (in Norwegian), Institute of Occupational Health, Oslo, 1978.

44. **Bonnet, J.**, Quantitative analysis of benzo(a)pyrene in vapors coming from melted tar, Natl. Cancer. Inst. Monogr. No. 9, 1962, 221.

45. **Sawicki, E., Fox, F. T., Elbert, W., Hauser, T. R., and Meeker, J.**, Polynuclear aromatic hydrocarbon composition of air polluted by coal tar pitch fumes, *Am. Ind. Hyg. Assoc. J.*, 23, 482l, 1962.

46. **Larson, B. A.**, Occupational exposure to coal tar pitch volatiles at pipeline protective coating operations, *Am. Ind. Hyg. Assoc. J.*, 39, 250, 1978.

47. **Gammage, R. B.**, Polycyclic aromatic hydrocarbons in work atmospheres, in *Handbook of Polycyclic Aromatic Hydrocarbons*, Bjørseth, A., Ed., Marcel Dekker, New York, 1983, 653.

48. **Cubit, D. A. and Tanita, R. K.**, Industrial Hygiene Assessment of Coal Gasification Plants, Final Summary Rep. by Environ. Control Division, Dynamac Corporation for the National Institute for Occupational Safety and Health, March, 1982.

49. **Bagchi, N. J. and Zimmerman, R. E.**, An industrial hygiene evaluation of chimney sweeping, *Am. Ind. Hyg. Assoc. J.*, 41, 297, 1980.

50. **Cadez, E.**, Ergebnisse langjähriger Untersuchunger über den Einfluss von Schadstoffen auf die Gesundheit von Schornsteinfegern, *Staub-Reinhalt. Luft*, 43, 116, 1983.

51. **Grimmer, G.**, Occurrence of PAH, in *Environmental Carcinogens: Polycyclic Aromatic Hydrocarbons*, Grimmer, G., Ed., CRC Press, Boca Raton, Fla., 1983, chap. 3.

52. **Lee, F. S.-C. and Schuetzle, D.**, Sampling, extraction, and analysis of polycyclic aromatic hydrocarbons from internal combustion engines, in *Handbook of Polycyclic Aromatic Hydrocarbons*, Bjørseth, A., Ed., Marcel Dekker, New York, 1983, chap. 2.

53. **Hakala, E., Anttonen, H., and Yrjänheikki, E.**, Polycyclic aromatic hydrocarbons in the mine atmosphere, in 30th Scandinavian Meeting on Occupational Hygiene, Åbo, Finland, 1981.

54. **Waller, R. E.**, Trends in lung cancer in London in relation to exposure to diesel fumes, in Symp. on Health Effects of Diesel Engine Emissions, Cincinnati, December, 1979.

55. **Commins, B. T., Walker, R. E., and Lawther, P. J.**, Air pollution in diesel bus garages, *Br. J. Ind. Med.*, 14, 232, 1957.

Chapter 9

PAH PROFILES AND PROXY METHODS

The ultimate goal for measuring PAH in workplace environments is the introduction of regulatory standards to control worker exposure to specific PAH compounds, or groups of these PAH compounds, that provide measures of health risk. For routine industrial hygiene monitoring purposes, rapid, simple, and low cost analytical methods are desired. Often, it seems sufficient to use a proxy method for PAH measurements, i.e., a method that simply measures one parameter or one compound that in some manner is representative of the very large and complex group of PAH compounds found in workplace samples. The use of proxy methods for workplace monitoring of PAH has recently been discussed.[1] Prerequisite for the selection of monitoring techniques and the choice of proxy compounds is a detailed and unbiased analysis of the working environment. It is of importance to first get a thorough knowledge of the relative composition of PAH in the samples and to determine the variation with time and with different plant locations. Only if the distribution pattern of PAH is relatively constant, may proxy methods be used to measure either one of the PAH compounds, or PAH as a group, to establish total PAH exposure.

The concept of parent PAH profile (PPP) has been utilized to characterize the PAH distribution and to establish baseline data for simplified PAH monitoring methods.[2] The idea behind the PPP is to illustrate the distribution pattern of a given number of key PAH compounds in the sample. The profile can be expressed as the observed concentrations[3] or, more commonly, as the relative distribution, e.g., as percent of total PAH[4], or normalized to a single PAH.[5]

The method has been applied to a number of samples from working atmospheres, as well as to PAH-containing materials used in industrial processes. Figure 1 shows the PAH profile measured in a Norwegian aluminum plant.[4] There are both vertical pin Söderberg and closed, prebaked cells in the primary aluminum smelter. Other plant operations associated with potential PAH exposure are in the anode baking plant. It has been postulated[4] that the PAH profile is characteristic of the process involved, i.e., the basic composition of the organic material, redox conditions, and process temperature. As revealed in the figure, the PAH profiles of the prebaked and the anode baking plant are relatively parallel. As these processes essentially use the same carbonaceous material at approximately the same temperatures, this supports the previously expressed idea that the profile is characteristic of the PAH source.

The Söderberg plant operation employs a process which is different from the process used at the other two plants. The PAH profile is accordingly different. In this case there is a significantly greater fraction of higher-boiling PAH present.

The profile found for the vertical pin Söderberg aluminum smelter is quite similar to the PAH profile found in a Norwegian coke plant.[5] This is attributed to the similarly, high operation temperatures of around 1000°C for both aluminum and coke production.

Figure 2 shows the PAH profiles for typical airborne samples taken from various work environments. The PPP for the iron and steel works roughly parallels those for coking and aluminum smelting. Notable differences exist in the cases of foundries and ferroalloy plants. The PAH profile for the foundry atmosphere is exceptional for the absence of benz(a)anthracene. To the best of our knowledge, the lack of benz(a)anthracene has not been confirmed in other studies. However, Schimberg et al.[6] have demonstrated that the distribution of PAH in emissions from foundry molding sands is strongly dependent on the type of hydrocarbon carrier used in the sand. Figure 3 shows PAH profiles for emissions from some typical additives used in molding sands. The proportion of higher PAH in emissions from coal powder is considerably larger than for coal tar pitch and starch.

The PAH profile for the electrode manufacturing department of a ferroalloy plant (Figure

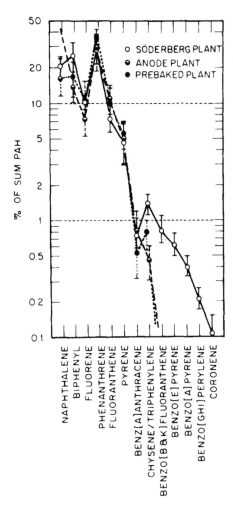

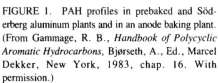

FIGURE 1.   PAH profiles in prebaked and Söderberg aluminum plants and in an anode baking plant. (From Gammage, R. B., *Handbook of Polycyclic Aromatic Hydrocarbons,* Bjørseth, A., Ed., Marcel Dekker, New York, 1983, chap. 16. With permission.)

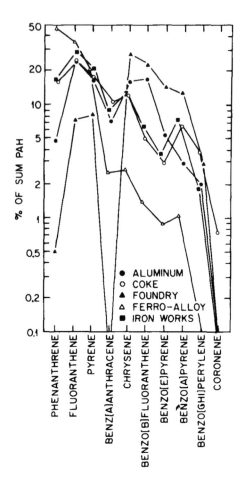

FIGURE 2.   PAH profiles for various work atmospheres. (From Gammage, R. B., *Handbook of Polycyclic Aromatic Hydrocarbons,* Bjørseth, A., Ed., Marcel Dekker, New York, 1983, chap. 16. With permission.)

2) is dominated by four compounds — phenanthrene, anthracene, fluoranthene, and pyrene. The proportion of higher-boiling PAH is in all cases less than 3% of the total amount. The lack of larger PAH is attributable to electrodes of anthracite and pitch-anthracene oil mixtures that are processed at the relatively low temperature of 120 to 190°C.

These examples demonstrate that great care should be exercised in the selection of specific PAH compounds that can act as proxies for a spectrum of PAH compounds. Without the knowledge of the PAH profiles and their variation at different plant locations, it becomes questionable to concentrate on a single proxy, such as BaP. From the nature and constancy of these profiles, proxy compounds can be chosen which will be indicative of the carcinogenic constituents.

It has been observed by Bjørseth et al.[7] that the profiles of PAH compounds in plants with one major PAH source are relatively constant for different samples. Blome and Baus[5a] found very similar PAH profiles at the battery top of four different coke plants (Figure 4). In these cases, monitoring by proxy may be used to indicate the exposure to total PAH. It

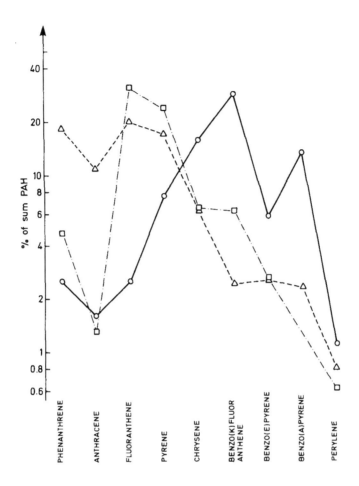

FIGURE 3. PAH profile of emissions from molding sands containing different carbonaceous materials.[6]

is reasonable to select a carcinogenic proxy PAH compound such as BaP, but it need not exhibit the carcinogenic activity directly; co-carcinogenic compounds such as pyrene and fluoranthene are active in enhancing the tumorigenic effects of BaP several-fold.[8] These PAH compounds are almost universally abundant as major constituents in real-life PAH mixtures. In contrast, BaP may be present only as a trace or minor constituent and, therefore, be difficult to detect. There are several types of samples that contain BaP at concentrations two orders of magnitude less than the most abundant PAH.[1] These include creosote, synthoil, and asphalt. For monitoring PAH exposure, the selection of BaP as an indicator would not be optimal in these cases.

Furthermore, if one is concerned with the exposure to vapors of PAH compounds, the very low equilibrium vapor concentration of BaP (77 ng/m^3 at $25°C)[9]$ makes it unsuitable as an indicator. The most appropriate proxy would be a volatile, low-boiling compound having a relatively high vapor pressure. Naphthalene or its methyl derivatives measured in real-time by second derivative UV spectroscopy, have been proposed as proxies to monitor PAH exposure from synthetic fuel production.[10]

In some work environments, the PAH profile of products and fugitive emissions may vary considerably at different locations within the plant. This seems to be true for coal-conversion plants. As quite different products are being processed or produced in the different areas of the plant, it is to be expected that fugitive emissions from different plant areas should have substantially different PAH profiles. Indeed, in a recent investigation of coal liquefaction

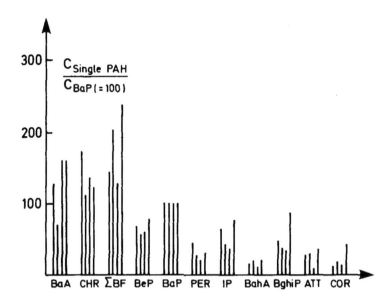

FIGURE 4.   PAH profile at the battery top of four different coke works. BaA, benz(a)anthracene; CHR, chrysene; ΣBF, benzo(b + j + k)fluoranthene; BeP, benzo(e)pyrene; BaP, benzo(a)pyrene; PER, perylene; IP, indeno(1,2,3-cd)pyrene; BahA, dibenzo(a,h)anthracene; BghiP, benzo(ghi)perylene; ATT, anthanthrene; COR, coronene. (From Blome, H. and Baus, K., Staub-Reihalt. Luft, 43, 367, 1983. With permission.)

plants, a striking variability in the relative proportions of PAH was found in air samples taken at different locations.[11] This feature makes the evaluation of long-term exposures of individual workers a difficult task. Furthermore, it does not seem likely that the characterization and monitoring procedures can be easily simplified by the use of a single proxy compound.

SUMMARY AND CONCLUSION

The relative distribution of PAH (PAH profile) in workplace atmospheres is characteristic of the process involved. For processes using similar carbonaceous materials and process conditions, similar PAH profiles are observed. In those cases where the nature of the profile has been investigated and its constancy has been proven, a proxy compound or compounds can be chosen which will be indicative of the concentration of all PAH. The proxy compound should be present in sufficient concentrations to allow a precise and accurate result, and it should be measurable by rapid and simple methods.

REFERENCES

1. **Gammage, R. B. and Bjørseth, A.,** Proxy methods and compounds for workplace monitoring of polynuclear aromatic hydrocarbons, in *Polynuclear Aromatic Hydrocarbons: Chemistry and Biological Effects,* Bjørseth, A. and Dennis, A. J., Eds., Battelle Press, Columbus, Ohio, 1980, 565.
2. **Bjørseth, A.,** Measurements of PAH content in workplace atmospheres, in *Luftverunreinigungen durch polycyclische aromatische Kohlenwasserstoffe — Erfassung und Bewertung,* VDI-Berichte 358, VDI-Verlag, Düsseldorf, 1980, 81.

3. **Bjøorseth, A. and Lunde, G.**, Analysis of the polycyclic aromatic hydrocarbon content of airborne particulate pollutants in a Söderberg paste plant, *Am. Ind. Hyg. Assoc. J.*, 38, 224, 1977.

4. **Bjørseth, A., Bjørseth, O., and Fjeldstad, P. E.**, Polycylic aromatic hydrocarbons in the work atmosphere. I. Determination in an aluminum plant, *Scand. J. Work Environ. Health*, 4, 212, 1978.

5. **Bjørseth, A., Bjøorseth, O., and Fjeldstad, P. E.**, Polycyclic aromatic hydrocarbons in the work atmosphere. II. Determination in a coke plant, *Scand. J. Work Environ. Health*, 4, 224, 1978.

5a. **Blome, H. and Baus, K.**, Konzentrationen polycyclischer aromatischer Kohlenwasserstoffe (PAH) bei Herstellung und Verwendung von Pyrolyseprodukten aus organischen Material, *Staub-Reinholt. Luft*, 43, 367, 1983.

6. **Schimberg, R. W., Toivonen, E., and Tossavainen, A.**, Polycyclische aromatische Kohlenwasserstoffe und andere Schadstoffe aus Giesseriformsanden mit verschiedenen Kohlenstoffträgern, *Staub-Reinhalt. Luft*, 41, 221, 1981.

7. **Bjørseth, A., Bjørseth, O., and Fjeldstad, P. E.**, Polycyclic aromatic hydrocarbons in the work atmosphere. Determination of area-specific concentrations and job-specific exposure in a vertical pin Söderberg aluminum plant, *Scand. J. Work Environ. Health*, 7, 223, 1981.

8. **Hoffmann, D., Schueltz, I., Hecht, S. S., and Wynder, E. L.**, Tobacco carcinogenesis, in *Polycyclic Hydrocarbons and Cancer*, Vol. 1, Gelboin, H. V. and Ts'o, P. O. P., Eds., Academic Press, New York, 1978, 85.

9. **Pupp, C., Lao, R. C., Murray, J. J., and Pottie, R. F.**, Equilibrium vapour concentrations of some polycyclic aromatic hydrocarbons, As_4O_6 and SeO_2 and the collection efficiencies of these air pollutants, *Atmos. Environ.*, 8, 915, 1974.

10. **Gammage, R. B., Vo-Dinh, T., Hawthorne, A. R., Thorngate, J. H., and Parkinson, W. W..**, A new generation of monitors for PAH from synthetic fuel production, in *Carcinogenesis — a Comprehensive Survey*, Vol. 3, Jones, P. W. and Freudenthal, R. I., Eds., Raven Press, New York, 1978, 155.

11. **Gammage, R. B.**, Polycyclic aromatic hydrocarbons in work atmospheres, in *Handbook of Polycyclic Aromatic Hydrocarbons*, Bjørseth, A., Ed., Marcel Dekker, New York, 1983, chap. 16.

Chapter 10

METHODS OF ANALYSIS

I. INTRODUCTION

For the determination of airborne PAH, there exist several alternative methods of sampling and numerous analytical methods. However, not all combinations of sampling and analysis are practicable, and not all the practicable combinations have been tried. Several important considerations are readily apparent. The first is that the method must be sensitive, accurate, and reproducible. Next, the method must be related to the health hazard. Third, the method should be as simple and economical as possible. Last, the method should not present undue hazards to laboratory personnel.

In selecting a method for the analysis of PAH in workplace atmosphere, one must choose from a wide variety of alternatives, ranging from the mere determination of the total weight of material collected on a filter, to the analysis of all identifiable components. This chapter does not pretend to give a complete description of all possible methods. Instead, we have chosen methods which are either widely used or which have been thoroughly checked.

The first method describes the determination of the cyclohexane- (or benzene-) extractable fraction of airborne particles as an index for the presence of PAH with carcinogenic properties. The second method is based on the fluorometric determination of BaP after TLC separation. In the third method, a complete profile analysis of PAH is given by CGC and FID. The fourth method describes the analysis of major PAH in both particle and gas phase by HPLC with fluorescence detection.

II. CYCLOHEXANE-SOLUBLE FRACTION OF TOTAL PARTICULATE MATTER

A. Scope and Field of Application

This method specifies a procedure for the determination of a selected fraction of airborne particles collected on filters, which is obtained by liquid extraction of components from the solid material. The cyclohexane-soluble fraction is used as an index for the presence of CTPV, including PAH.

Several solvents can be used for the extraction of CTPV from particles.[1] Benzene has been widely used as a solvent in extractions of this type, resulting in the BSF. Because of the more recent awareness of the toxic properties of benzene,[2] cyclohexane has been recommended as a substitute for benzene in the extraction of CTPV from particulate matter.[3]

The benzene- and cyclohexane-soluble fractions have been used as a method of controlling workers' exposure to CTPV in a large number of different workplace atmospheres. These include coke ovens,[4-6] aluminum plants,[7-9] iron and steel works,[10] asphalt production,[11] and chimney sweeping.[12]

B. Principle

The cyclohexane-soluble material in the particles collected on a glass fiber filter is extracted by ultrasonication. After extraction, the cyclohexane solution is filtered through a fritted glass funnel. An aliquot of the extract is carefully evaporated to dryness and weighed. Blank filters are extracted in the same manner as the samples.

C. Apparatus

Apparatus used are usual laboratory equipment and the following items:

1.   Ultrasonic bath, partially filled with water
2.   Electrobalance capable of weighing to 1 μg
3.   Teflon weighing cups, 2 mℓ, approximate tare weight 60 mg
4.   Glass fiber filters, 37 mm diameter, Gelman Type A or equivalent
5.   Silver membrane filters, 37 mm diameter, 0.8 μm pore size
6.   Vacuum oven
7.   Glassine paper, 9 × 12 cm
8.   Funnels, fritted glass, 15 mℓ
9.   Graduated evaporative concentrator, 10 mℓ

D. Reagents

1.   Cyclohexane, ACS nanograde reagent
2.   Dichromic acid cleaning solution
3.   Acetone, ACS reagent grade

E. Procedure

1.   All extraction glassware is cleaned with dichromic acid cleaning solution, rinsed with tap water, then with deionized water followed by acetone, and allowed to dry completely. The glassware is rinsed with nanograde cyclohexane before use. The Teflon cups are cleaned with cyclohexane and acetone.
2.   The Teflon cups are preweighed to 0.01 mg.
3.   Remove top of cassette and hold over glassine paper. Remove plug on bottom of cassette. Insert end of application stick through hole and gently raise filters to one side. Use tweezers to remove filters, and loosely roll filters around tweezers. Slide rolled filters into test tube and push them to bottom of tube with application stick. Add any particles remaining in cassette and on glassine paper to test tube.
4.   Pipet 5 mℓ cyclohexane into test tube 150 × 16 mm.
5.   Put test tube into a 50-mℓ beaker filled with water above level of cyclohexane in the test tube. Put beaker into the ultrasonic bath.
6.   Sonify sample for 5 min.
7.   Filter the extract in 15 mℓ medium fritted glass funnels.
8.   Rinse test tube and filters with two 1,5-mℓ aliquots of cyclohexane, and filter through the fritted glass funnel.
9.   Collect the extract and two rinses in the 10-mℓ graduated evaporative concentrator.
10. Evaporate down to 1 mℓ while rinsing the sides with cyclohexane.
11. Pipet 0.5 mℓ of the extract to preweighed Teflon weighing cup.
12. Evaporate to dryness in a vacuum oven at 40°C for 3 hr.
13. Allow the weighing cups to equilibrate in the balance room for $^{1}/_{2}$ to 1 hr before weighing. Use counterweighing techniques on electrobalance with full scale range of 1 mg to determine weight aliquot to nearest microgram.

F. Calculations
   The amount of cyclohexane-soluble fraction present in the sample (in milligrams) may be determined according to the following equation:

$$\text{mg/sample} = 2 \times [\text{wt sample aliquot (mg)} - \text{wt blank aliquot (mg)}]$$

The amount of cyclohexane-extractable fraction present in the air may be determined according to the following equation:

$$mg/m^3 = \frac{mg/sample}{air\ volume\ collected\ (m^3)}$$

G. Repeatability and Reproducibility

The SD for 9 analyses of a benzene solution containing 1350 µg benzene solubles was 2%.[3] Other workers[13] state SD of less than 3% for the determination of cyclohexane-soluble matter in aluminum plants.

Collaborative studies of benzene solubles have been performed by five laboratories within the aluminum industry. SD were 0.9 to 4.0% at the 0.77 to 1.26 mg/mℓ level.[14]

H. Schematic Presentation of the Procedure

Sample filter

↓

Extract with cyclohexane by ultrasonication

↓

Filter extract

↓

Concentrate to 1 mℓ

↓

Evaporate aliquot in weighing cups

↓

Determine weight gain to nearest microgram

I. Advantages and Disadvantages

Advantages — The method is simple and fast and gives good precision. It does not require sophisticated or expensive equipment. It requires only a minimum of technician training.

Disadvantages — The major deficiency of the method is that it is nonspecific for PAH and is subject to a variety of interferences. Often the workplace atmosphere contains aliphatic lubricating oils, combustion oils, or other contaminants which are soluble in cyclohexane or benzene. The amount of these less-hazardous non-PAH solubles may vary considerably. Their presence will overstate the potential hazard arising from the measurement of the cyclohexane- or benzene-soluble fraction.

III. ANALYSIS OF BaP

A. Scope and Field of Application

This method specifies a procedure for the determination of BaP in airborne particles sampled on glass fiber filters.[15] The method is suitable for air samples from occupational environments in which the amount of BaP ranges from about 1 ng/m³ to above 250 µg/m³. Both stationary and personal sampling may be used. The method has been used extensively for the determination of the exposure to PAH in a great number of work environments in Sweden.[16]

B. Principle

PAH are desorbed from the filters by vacuum sublimation. The condensate is dissolved in cyclohexane (or cyclohexane/aceton) and aliquots are submitted to TLC on acetylated cellulose. BaP is quantitated by fluorescence scanning of the TLC plate. The recovery is estimated by an isotope dilution technique.

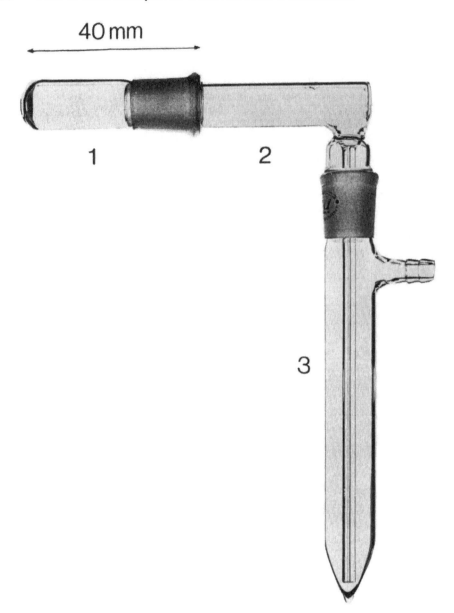

FIGURE 1. Vacuum sublimation apparatus consisting of a test tube (1), condenser (2), and receiver (3).[15]

C. Apparatus

Apparatus used are usual laboratory equipment and the following items:

1. The all-glass vacuum sublimation apparatus, as shown in Figure 1; it consists of a test tube (1), condenser (2), and receiver (3)
2. Rotary vacuum pump with an ultimate vacuum of at least 1 Pa
3. Metal block thermostat fitting to test tube of vacuum sublimation apparatus
4. Precoated thin-layer plates of 30% acetylated cellulose 0.1 mm, 20 × 20 cm
5. Developing tanks
6. Thin-layer applicator
7. Spectrofluorometer with scanner for direct reading from LC plate; the fluorometer may

have either 2 monochromators (for excitation and emission) or a 366 nm filter in the excitation path and a monochromator in the emission path
8.   Recorder and area integrator
9.   10-$\mu\ell$ syringe
10. Liquid scintillation spectrometer

D. Reagents

1.   Glass-distilled cyclohexane
2.   Glass-stilled acetone
3.   Ethanol, 99% analytical grade
4.   Dichloromethane, analytical grade
5.   BaP and (G-^{3}H) BaP
6.   Scintillation solution

E. Procedure

Note: BaP is degraded by UV light and exposure of samples, extracts, or standard solutions to sources such as sunlight should be avoided. All operations should be carried out under subdued light.

1.   Place glass fiber filter into the upper, horizontal tube of the vacuum sublimation apparatus (Figure 1). Assemble the apparatus and connect vacuum pump to obtain a vacuum below 10 Pa.
2.   Insert the test tube end into the metal block thermostat maintained at 250°C and leave it for 1 hr.
3.   Release vacuum and cool the apparatus. Remove test tube. Turn apparatus upside down and introduce 5 mℓ cyclohexane into the condenser (Figure 1). If the condensate has not dissolved completely in cyclohexane, add acetone until complete dissolution. The solution is allowed to run into the receiver.
4.   Remove the condenser and evaporate to dryness under a gentle stream of nitrogen.
5.   Dissolve the residue in a minimal amount of cyclohexane (>20 $\mu\ell$) or cyclohexane/ acetone, 2:1.
6.   Apply 2 parallels of 2 $\mu\ell$ from 2 samples on the thin-layer plate. Apply 2 $\mu\ell$ of each of 3 standard solutions (between 0.25 and 10 ng BaP per $\mu\ell$ on the same TLC plate.
7.   Place the TLC plate into the developing tank containing ethanol/dichloromethane/ water, 20:10:1. When the solvent front has migrated to the 15 cm mark (about 2 hr), remove the TLC plate and let it dry in a hood.
8.   The fluorometric determination of BaP is made directly on the TLC plate by means of a fluorescence scanning densitometer. The excitation wavelength is set to 366 nm and the emitted fluorescence light is recorded at 407 nm.
9.   Scanning is made in the running direction of each spot. The BaP peak of the samples is identified in comparison to the coordinates on the plate with the reference spot.
10. The calculation of the amount BaP is made from peak areas. Every TLC plate is treated as a separate assay.
11. For the estimation of recovery, spread 200 $\mu\ell$ of a solution of ^{3}H-BaP, corresponding to 15 to 20 nCi, over a sample filter and let it dry. Follow the procedure from steps 1 through 5, and take out aliquots for liquid scintillation counting. Follow instructions for liquid scintillation spectrometer. Calculate recovery from the radioactivity measured.

F. Calculation

1. Determine amount of BaP in a sample in ng per sample by comparing areas for the sample and standards on the same TLC plate.
2. If necessary, correct for recovery determined by isotope dilution.
3. The air concentration is calculated by dividing the amount of BaP (nanograms per sample) by the total volume of air sampled.

G. Repeatability and Reproducibility

The coefficient of variation of the analytical procedure is estimated to be 10% in the range 10 ng to 2.5 µg BaP per filter. The reproducibility of the method is affected by several variables such as light-induced oxidation, uniformity of the coating on the TLC plates, the technique of spotting the plates, and other factors.

H. Schematic Representation of the Procedure

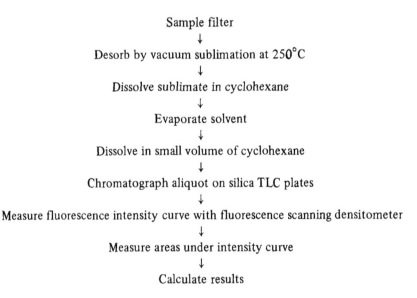

Sample filter
↓
Desorb by vacuum sublimation at 250°C
↓
Dissolve sublimate in cyclohexane
↓
Evaporate solvent
↓
Dissolve in small volume of cyclohexane
↓
Chromatograph aliquot on silica TLC plates
↓
Measure fluorescence intensity curve with fluorescence scanning densitometer
↓
Measure areas under intensity curve
↓
Calculate results

Control recovery by spiking a sample filter with ³H-BaP and determining radioactivity of the condensate solution.

I. Advantages and Disadvantages

Advantages — The method is rapid and very sensitive. There is no repeated handling and transferring of the sample unknown.

Disadvantages — Only BaP is determined as a representative of the very large and complex groups of PAH compounds present in samples from workplace environments. It has been shown that the PAH profile may vary greatly between different industries and at different plant locations.[17] BaP may be present only as a trace amount or minor constituent. Thus the amount of BaP measured may not be an adequate indicator of the total of PAH present, and may not reflect the carcinogenic potential of the exposure.

IV. GAS CHROMATOGRAPHIC PROFILE ANALYSIS OF PAH

A. Scope and Field of Application

This method specifies a gas chromatographic procedure for the determination of the quantitive distribution of PAH in samples from workplace atmospheres.[17,18] The method can determine nanogram quantities of the various PAH. It has been used extensively in establishing PAH profiles in a number of industrial atmospheres in Norway and in monitoring exposure to PAH.[17]

B. Principle

The PAH collected on the filter are extracted with cyclohexane containing two internal standards. For clean-up, the PAH are partitioned into DMF/water (9:1). After addition of water and extraction into cyclohexane, a final clean-up on silica gel is performed.

Individual PAH are determined by CGC with (FID). Results are calculated by comparison with the area of FID signals of the internal standards.

C. Apparatus

All glassware is thoroughly rinsed and finally cleaned by heating to 550°C. Apparatus used are usual laboratory equipment and the following items:

1. Soxhlet extraction apparatus, 40 mℓ
2. Modified Kuderna-Danish evaporator (see Figure 2)
3. Glass column for silica chromatography, 10 mm I.D. $\times$ 20 mm with solvent reservoir
4. Gas chromatograph: any instrument for CGC equipment with splitless injector (or on-column injector), FID, recorder, and electronic area integrator
5. Wall-coated open tubular (WCOT) 25 m $\times$ 0.25 to 0.35 mm I.D. glass or fused silica capillary column, persilylated or polysiloxane deactivated, coated with a 0.05 to 0.15 μm film of SE-54 (perferably immobilized)

D. Reagents

1. Cyclohexane, analytical grade, redistilled in glass vessels
2. Dimethylformamide, analytical grade, vacuum-distilled twice in glass vessels
3. Silica gel, for adsorption chromatography, e.g., Woelm Pharma, West Germany, 70 to 150 mesh deactivated with 15% water 24 hr before use
4. Na_2SO_4, heated to 450°C

E. Procedure

Note: PAH are degraded by UV light and exposure of samples, extracts, or standard solutions to sources such as sunlight should be avoided. To ensure purity of reagents and glassware used, a blank test should be carried out using the following procedure, but omitting the samples.

Extraction

1. Weighed samples of glass fiber filters are placed in the extraction thimble of the Soxhlet apparatus.
2. Add 60 mℓ cyclohexane together with 2 to 10 μg of 3,6-dimethylphenanthrene and 2,2'-binaphthyl (20 to 100 $\mu\ell$ of standard solution) and reflux overnight (15 hr).
3. Allow the organic extract to cool and transfer it to a 250 mℓ separation funnel.

Liquid-Liquid Partition Clean-Up

1. Add 50 mℓ of freshly prepared DMF/water, 9:1 and shake carefully to avoid formation of emulsions. Allow to separate.

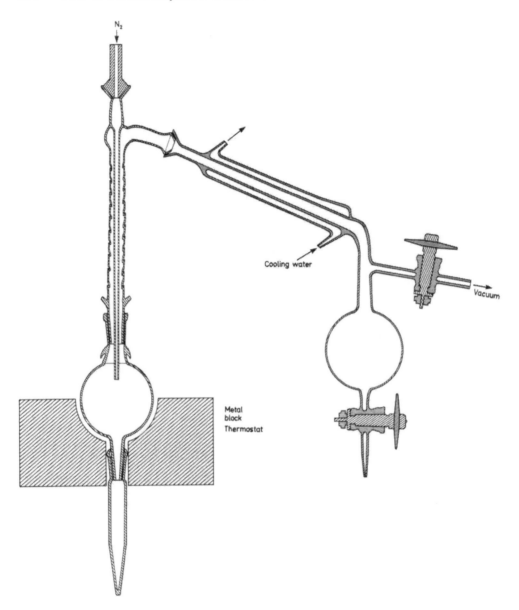

FIGURE 2.   Modified Kuderna-Danish evaporator for concentrating PAH extracts under nitrogen at reduced pressure.

2.   Transfer upper layer (cyclohexane) to another 250 mℓ separation funnnel and extract once more with 25 mℓ of DMF/water, 9:1.
3.   Add 75 mℓ of water to the combined DMF/water layers and allow to cool to room temperature.
4.   Extract DMF to water phase, first with 50 mℓ then with 25 mℓ of cyclohexane.
5.   Combine the cyclohexane layers (upper) and wash 2 times with 25 mℓ water. Dry cyclohexane phase with Na_2SO_4 for at least 2 hr.
6.   Concentrate cyclohexane extract to ca. 8 mℓ by use of a modified Kuderna-Danish apparatus (Figure 2) under nitrogen at reduced pressure. Keep temperature below 50°C and the pressure above 1.3×10^3 Pa to ensure minimum loss and decomposition of PAH during the concentration step.

7.   Concentrate further to about 2 mℓ in a centrifuge tube at 30°C by directing a gentle stream of highly purified nitrogen onto the liquid surface.

Column Chromatography Clean-Up

1.   To prepare the column, add 20 mℓ cyclohexane to 5 g silica gel. Shake to degas, then add slurry to the glass column and run off solvent until liquid level coincides with the upper surface of the silica gel.
2.   Transfer concentrated extract to top of column. Rinse centrifuge tube with 1 mℓ cyclohexane and add to the same column. Run off solvent until the liquid level coincides
3.   Elute the column with 100 mℓ cyclohexane.
4.   Concentrate eluate by use of a modified Kuderna-Danish apparatus under nitrogen and reduced pressure.
5.   Concentrate further to between 0.1 and 0.5 mℓ in a centrifuge tube at 30°C under a gentle stream of nitrogen.

Gas Chromatographic Analysis

1.   Gas chromatographic conditions should be as follows:
     column: WCOT 25 m × 0.35 mm I.D. glass or fused silica capillary coated with a 0.15 μm film of SE-54 or similar; **carrier gas:** H_2 or He, linear velocity approximately 50 cm/sec or 30 cm/sec, respectively; **injector temperature:** 250°C; and **detector temperature:** 275°C.
2.   Inject 2 μℓ of concentrated sample splitless at an oven temperature of 60°C using a 10-μℓ syringe. Open split after about 40 sec.
3.   The column is heated at a high rate (about 30°C/min) to 120°C, 4 min after injection, followed by a temperature program of 3° C/min to a final temperature of 300°C.

Characterization of the PAH Signals

Using two internal standards (3,6-dimethylphenanthrene and 2,2'-binaphthyl) which are easy to detect by the size of the signals, the characterization of a PAH signal presents no difficulties. Relative retention times of peaks in the chromatogram of an unknown sample may be compared with those of standards or with those identified by GC/MS.

Quantification

Results are calculated by comparing the electronically recorded areas of the FID signals of each PAH with the area of the nearest internal standard. Usually a correction for different relative response factors of the various PAH is not necessary. Routinely, about 40 polycyclic aromatic compounds are quantitated. Table 1 shows an example of the relative distribution of PAH sampled on filter in a Söderberg aluminum plant. Compounds above 0.1% are shown, which constitute 32 of a total of 39 PAH determined.

F. Calculation

The amount, W_x, of PAH, x, in the sample is given by

$$W_x = \frac{A_x \times W_i}{A_i}$$

where A_x = peak area of PAH x, A_i = peak area of respective internal standard added, and W_i = weight of respective internal standard added (μg). The concentration of PAH in air is calculated by dividing the sum of the amount of all identified PAH (μg/sample) by the total volume of air sampled.

Table 1
RELATIVE DISTRIBUTION OF PAH ON FILTER
FROM A SÖDERBERG ALUMINUM PLANT

PAH	Relative distribution
Acenaphthene	0.1
Dibenzofuran	0.2
Fluorene	0.3
2-Methylfluorene	0.1
Dibenzothiophene	0.6
Phenanthrene	10.7
Anthracene	3.1
Carbazol	2.7
3-Methylphenanthrene	2.1
2-Methylphenanthrene	2.2
2-Methylanthracene	1.3
4,5-Methylenephenanthrene	2.3
4-and/or 9-Methylphenanthrene	1.3
1-Methylphenanthrene	1.5
Fluoranthene	25.6
Pyrene	17.0
Benzo(a)fluorene	5.6
Benzo(b)fluorene	3.2
1-Methylpyrene	1.5
BaA	5.0
Chrysene and triphenylene	4.6
Benzo(b)fluoranthene	1.3
Benzo(j and K)fluoranthene	1.3
BeP	1.2
BaP	0.9
Perylene	0.6
Indeno(1,2,3-cd)pyrene	0.9
Dibenz(a,c and/or a,h)anthracene	0.6
Benzo(ghi)perylene	0.8
Anthanthrene	0.6
Dibenzo(a,e)pyrene	0.3
Coronene	0.3

From Fjeldstad, P. E., *Polycyclic Aromatic Hydrocarbons in the Primary Aluminum Industry,* Nordheim, E. and Guthe, T., Eds., Nordic Aluminum Industry's Secretariate for Health, Environment and Safety, Oslo, Norway, 1984. With permission.

G. Repeatability and Reproducibility

The repeatability of the method was tested by analysis of a set of five equivalent samples of particulate matter from an aluminum reduction plant. For 15 compounds which represented both major and minor components, the SD ranged from 1.6 to 9.7%, with an average of 4.5%.[18]

A round robin test of PAH analysis in particulate matter was performed among laboratories in the Nordic countries.[19] For 7 laboratories using the described method, the mean RSD for the sum of 8 chosen PAH was 16%. The RSD for the individual PAH, however, varied from 16 to 46%.

H. Schematic Representation of the Procedure

Sample filter
↓
Extract in Soxhlet with cyclohexane
↓
Extract twice with DMF/water, 9:1, discard cyclohexane layer
↓
Add water and extract twice with cyclohexane, discard DMF/water layer
↓
Wash twice with water, discard water
↓
Concentrate to about 2 mℓ
↓
Transfer to silica column
↓
Elute with 100 mℓ cyclohexane
↓
Concentrate to 0.1 to 0.5 mℓ
↓
Analyze PAH by CGC with FID

I. Advantages and Disadvantages

Advantages — Several advantages can be derived from using CGC for analysis of PAH: (1) high separation efficiency, (2) good resolution among structural isomers, (3) high sensitivity, (4) good reproducibility, and (5) possibility for identification by combination with mass spectrometry.

Disadvantages — The method requires extensive sample clean-up prior to analysis.

V. HPLC DETERMINATION OF PAH

A. Scope and Field of Application

The method specifies a procedure for the determination of both gaseous and particulate PAH in workplace atmospheres. Various modifications of the method have been applied to ambient PAH[20] and to PAH in working environments.[21]

B. Principle

PAH are sampled on a filter with a back-up section of Amberlite XAD-2 adsorbent. Filter and adsorbent are solvent extracted and the extracts submitted to reverse-phase HPLC with fluorescence detection.

C. Apparatus

Apparatus used are usual laboratory equipment and the following items:

1. Sampling device consisting of a glass fiber filter (or PTFE membrane) in a standard 37-mm filter cassette equipped with 2 middle sections to accommodate about 3 g XAD-2 between filter and the support pad. The arrangement is shown in Figure 2, Chapter 5
2. Soxhlet extraction apparatus
3. HPLC column, 4.6 mm I.D. × 250 mm filled with polymeric C_{18} modified silica quality controlled for PAH analyses (e.g., Vydac 201 TP), 5 μm particle size

4.   Liquid chromatograph with gradient capability and spectrofluorometric detector
5.   Recorder and electronic integrator

D. Reagents

1.   Water — deionized, distilled, degassed
2.   Acetonitrile, HPLC grade, degassed
3.   PAH reference standards: naphthalene, acenaphthene, fluorene, phenanthrene, anthracene, fluoranthene, pyrene, BaA, chrysene, BeP, benzo(b)fluoranthene, benzo(k)fluoranthene, BaP, dibenz(a,h)anthracene, benzo(ghi)perylene
4.   Calibration stock solution, 0.25 mg PAH per mℓ acetonitrile; stable about 6 months if refrigerated and protected from light
5.   Amberlite XAD-2 purified by repeated washing with distilled water; remove fines by decanting; wash 5 times with methanol and extract in a Soxhlet apparatus for 12 hr with methylenechloride

E. Procedure
Note: This method has been developed for a wide range of sample types. Specific sample sets may require modification in extraction solvent used, choice of analytes, or choice of detection wavelengths.[22] All glassware used for the analysis should be washed with detergent and then thoroughly rinsed with tap and distilled water followed by distilled acetone.

Extraction

1.   Remove the filter and the XAD-2 resin from the sampling device and place them in the extraction thimble of separate Soxhlet extractors.
2.   Add methylenechloride and carry out extraction for 8 hr.
3.   Reduce volume in a Kuderna-Danish evaporator to 1 mℓ. Add ca. 1 mℓ acetonitrile, take to near dryness, and adjust final volume to 1 mℓ with acetonitrile.
4.   Filter all samples through 0.45 μm PTFE filters.
5.   If necessary, spectral interference from other compounds in the sample may be reduced by a sample fractionation step. Reduce the dichlormethane extract volume to 0.5 mℓ and apply to a 1 × 25 cm column of silica gel. Elute with 25 mℓ n-pentane followed by 25 mℓ of 40% CH_2Cl_2 in n-pentane. Collect the eluent of the last 25 mℓ and reduce its volume by solvent evaporation. After addition of 1 mℓ of acetonitrile, evaporate the remainder of the CH_2Cl_2 and n-pentane.

HPLC Analysis

Analyze the sample extracts by HPLC using the following conditions:

1.   Mobile-phase flow rate: 1.0 mℓ/min.
2.   Injection volume: 5 to 25 μℓ.
3.   Temperature: ambient. If room air is not at a constant temperature and draft-free, the column should be jacketed and kept at 25°C.
4.   Mobile phase: linear solvent gradient from 50% acetonitrile in water to 100% acetonitrile in 50 min. Hold 15 min at 100% acetonitrile. Return to initial conditions and equilibrate column for 10 min before next injection.
5.   Detection: spectrofluorometer wavelengths for 2- and 3-ring PAH, excitation 280 nm, emission 340 nm; for quantitation of 4-ring and larger PAH, excitation 305 nm, emission 430 nm.[23]

Calibration and Quality Control

1. Prepare working standards daily by diluting the calibration stock solution in 10 mℓ volumetric flasks, e.g., 2.5, 0.5, 0.1, 0.02, and 0.002 $\mu g/m\ell$.
2. Intersperse working standards and samples in the analysis.
3. Measure peak areas of standards. Prepare a calibration curve plotting peak areas vs. amount (ng) injected.
4. Determine desorption efficiencies for filter and adsorbent at least once for each batch of filters and adsorbents used in the sampling range of interest. Spike four filters at each five concentration levels with a mixture of PAH to be analyzed. Allow the filter to dry in the dark overnight. Analyze filters and plot desorption efficiency vs. amount found. Spike calibration solution at five concentration levels directly on adsorbent in suitable vials. Run four replicates at each concentration. Cap vials and allow to stand overnight. Analyze and plot desorption efficiency vs. amount found.
5. Analyze at least one field blank for each sample medium.

F. Calculations

Read the amount of analyte injected from the calibration curve. If analysis of a sample extract yields X_i μg injected of the *i*th PAH, the concentration of this compound in air sampled on either filter (C_f) or adsorbent (C_a) is given by

$$C_f = \frac{(X_i - B_i) \times 10^3}{f \times V \times DE_i}$$

$$C_a = \frac{(X_i - B_i) \times 10^3}{f \times V \times DE_i}$$

where B is the amount found in the analysis of blanks, f is the ratio of the injection volume to the total extract volume, V is the volume (in ℓ) of the actual air volume sampled, and DE_i is the extraction efficiency for that PAH.

Note: C_a includes particulate analytes originally collected on filter, then volatilized during sampling, which is a significant fraction for many PAH (e.g., naphthalene, acenaphthene, fluorene, etc.).

G. Repeatability and Reproducibility

The reproducibility for the analysis of standard solutions of 13 PAH using HPLC, internal standards, and programed wavelength fluorescence detection was 1.5 to 2.5% RSD during nine replicate analyses. This level of uncertainty was found to be similar to that found when measuring peak areas with UV detection (254 nm) and fluorescence detection with constant excitation and emission wavelengths.[20] No data are available on the reproducibility of the method.

H. Schematic Presentation of the Procedure

Sample filter + adsorbent
↓
Soxhlet extract separately with CH_2Cl_2
↓
Concentrate to ca. 1 mℓ in Kuderna-Danish evaporator
↓
Exchange solvent to CH_3CN
↓
Filter through 0.45 μm PTFE filter
↓
Analyze by reversed-phase HPLC using fluorescence detection

I. Advantages and Disadvantages

Advantages — HPLC offers several advantages for the determination of PAH in complex mixtures. A variety of stationary phases are available that provide differing selectivities for PAH isomers. By proper choice of stationary and mobile phases, PAH isomers which generally are not separated by GC (e.g., chrysene/triphenylene and the benzofluoranthenes) can be separated easily by LC.[22] Fluorescence spectroscopic detection provides both a sensitive and selective means for monitoring the LC effluent.[24,25] Due to the unique selectivity of the detection, little or no clean-up of the extracts is necessary prior to analysis.

Disadvantages — A major disadvantage of HPLC is its lower separation efficiency as compared to CGC. Thus, a particular PAH may co-elute with the alkyl-substituted homologue of a PAH with a lower number of aromatic carbons.

As the number of PAH of analytical interest in increased, the chances of them having a common excitation and emission wavelength is reduced. Thus compromise wavelength pairs have to be chosen, thereby reducing the sensitivity and selectivity of the detection.

REFERENCES

1. **Sawicki, E.,** The separation and analysis of polynuclear aromatic hydrocarbons present in the human environment, I, II, and III, *Chemist-Analyst,* 53, 24, 1964.
2. International Agency for Research on Cancer, *IARC Monographs on the Evaluation of the Carcinogenic Risk of Chemicals to Humans,* Vol. 7, Some Anti-Thyroid and Related Substances, Nitrofurans, and Industrial Chemicals, IARC, Lyon, France, 1974, 203.
3. *NIOSH Criteria for a Recommended Standard...Occupational Exposure to Coal Tar Products,* U.S. Department of Health and Human Services (NIOSH), Publ. No. 78—107, 1977.
4. **Fannick, N., Gonshor, L. T., and Shockley, J., Jr.,** Exposure to coal tar pitch volatiles at coke ovens, *Am. Ind. Hyg. Assoc. J.,* 33, 461, 1972.
5. **Jackson, J. O., Warner, P. O., and Mooney, T. F., Jr.,** Profiles of benzo(a)pyrene and coal tar pitch volatiles at and in the immediate vicinity of a coke oven battery, *Am. Ind. Hyg. Assoc. J.,* 35, 276, 1974.
6. National Institute for Occupational Safety and Health, NIOSH Technical Information, Report on Analytical Methods Used in a Coke Oven Effluent Study, Publ. No. 74—105, National Institute for Occupational Safety and Health, U.S. Department of Health, Education and Welfare, 1974.
7. **Shuler, P. J. and Bierbaum, P. J.,** Environmental Survey of Aluminum Reduction Plants, National Institute for Occupational Safety and Health, Publ. No. 74—101, U.S. Department of Health, Education and Welfare, Cincinnati, 1974.
8. **Steinegger, A. F.,** PAH: Swiss experience, in *Health Protection in Primary Aluminium Production,* Vol. 2, Hughes, J. P., Ed., International Primary Aluminium Institute, London, 1981, 129.
9. **Bonney, T. B.,** PAH: North American experience, in *Health Protections in Primary Aluminium Production,* Vol. 2, Hughes, J. P., Ed., International Primary Aluminium Institute, London, 1981, 133.

10. **Tanimura, H.,** Benzo(a)pyrene in an iron and steel works, *Arch. Environ. Health,* 17, 172, 1968.

11. **Adamziak-Ziemba, J. and Kesy-Dabrowska, I.,** Pollution of air with benzene soluble substances and polycyclic aromatic hydrocarbons at asphalt production (in Polish), *Med. Pr.,* 23, 617, 1972.

12. **Bagchi, N. J. and Zimmerman, R. E.,** An industrial hygiene evaluation of chimney sweeping, *Am. Ind. Hyg. Assoc. J.,* 41, 297, 1980.

13. **Steinegger, A. and Glaus, R.,** The determination of polycyclic aromatic hydrocarbons at the workplace in carbon plant and reduction plant, *Light Met. (Warrendale, Pa),* 977, 1981.

14. **Seim, H. J.,** Kaiser Aluminium and Chemical Corporation, Pleasanton, Calif., private communication.

15. **Sollenberg, J.,** A method for determining benzo(a)pyrene in air samples collected on glass fiber filters in occupational areas, *Scand. J. Work Environ. Health,* 3, 85, 1976.

16. **Lindstedt, G. and Sollenberg, J.,** Polycyclic aromatic hydrocarbons in the occupational environment (in Swedish), *Arbete och Hälsa,* 1980:1, Arbetarskyddsverket, Stockholm, 1980.

17. **Bjørseth, A,.** Measurement of PAH content in workplace atmospheres, in *Luftverunreinigungen durch polycyclische aromatische Kohlenwasserstoffe,* VDI-Berichte 358, VDI-Verlag GmbH, Düsseldorf, 1980, 81.

18. **Bjørseth, A.,** Analysis of polycyclic aromatic hydrocarbons in particulate matter by glass capillary gas chromatography, *Anal. Chim. Acta,* 94, 21, 1977.

19. **Bjørseth, A. and Olufsen, B.,** Results from a Scandinavian Round Robin Test of PAH Analysis (in Norwegian), Nordic PAH Proj. Rep. No. 1, Central Institute for Industrial Research, Oslo, 1978.

20. **May, W. E. and Wise, S. A.,** Liquid chromatographic determination of polycyclic aromatic hydrocarbons in air particulate extracts, *Anal Chem.,* 56, 225, 1984.

21. **Andersson, K., Levin, J.-O., and Nilsson, C.-A.,** Sampling and analysis of particulate and gaseous polycyclic aromatic hydrocarbons from coal tar sources in the working environment, *Chemosphere,* 12, 197, 1983.

22. **Wise, S. A., Bonnett, W. J., Guenther, F. R., and May, W. E,** A relationship between reversed-phase C_{18} liquid chromatographic retention and the shape of polycyclic aromatic hydrocarbons, *J. Chromatogr. Sci.,* 19, 457, 1981.

23. **Ogan, K., Katz, A., and Slavin, W.,** Determination of polycyclic aromatic hydrocarbons in aqueous samples by reversed-phase liquid chromatography, *Anal. Chem.,* 51, 1315, 1979.

24. **Wise, S. A.,** High-performance liquid chromatography for the determination of polycyclic aromatic hydrocarbons, in *Handbook of Polycyclic Aromatic Hydrocarbons,* Bjørseth, A., Ed., Marcel Dekker, New York, 1983, chap. 5.

25. **Ogan, K., Katz, E., and Porro, T. J.,** The role of spectral selectivity in fluorescence detection for liquid chromatography, *J. Chromatogr. Sci.,* 17, 597, 1979.

INDEX

Printed and bound by CPI Group (UK) Ltd, Croydon, CR0 4YY

24/10/2024

01778897-0001